Port Newark and the Origins of Container Shipping

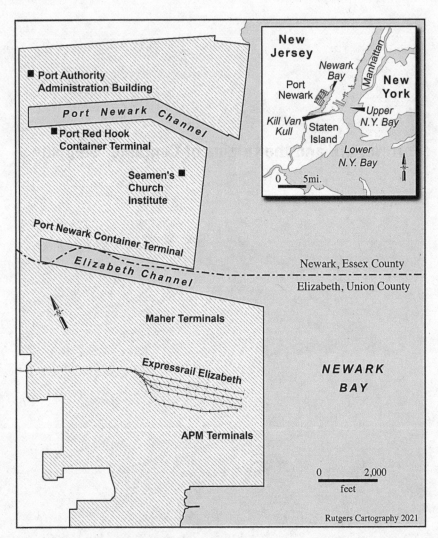

Map of Port Newark by Michael Siegel.

Port Newark and the Origins of Container Shipping

● ● ● ● ● ● ● ● ● ● ● ● ● ● ● ● ●

ANGUS KRESS GILLESPIE
Foreword by Michael Aaron Rockland

Rutgers University Press
New Brunswick, Camden, and Newark, New Jersey, and London

Library of Congress Cataloging-in-Publication Data

Names: Gillespie, Angus K., 1942- author.
Title: Port Newark and the origins of container shipping / Angus Kress Gillespie ;
   foreword by Michael Aaron Rockland.
Description: New Brunswick : Rutgers University Press, [2023] |
   Includes bibliographical references and index.
Identifiers: LCCN 2022007402 | ISBN 9781978818712 (cloth) |
   ISBN 9781978818729 (epub) | ISBN 9781978818743 (pdf)
Subjects: LCSH: Containerization—History. | New York Harbor (N.Y. and N.J.)—
   History. | Harbors—New Jersey—Newark—Management—History. | Shipping—
   New Jersey—Newark—History. | Shipping—New York (State)—New York—History. |
   McLean, Malcolm, 1914-2001.
Classification: LCC TA1215 .G55 2023 | DDC 627/.3097493—dc23/eng/20220503
LC record available at https://lccn.loc.gov/2022007402

A British Cataloging-in-Publication record for this book is available from the British Library.

Copyright © 2023 by Angus Kress Gillespie
All rights reserved

No part of this book may be reproduced or utilized in any form or by any means, electronic or mechanical, or by any information storage and retrieval system, without written permission from the publisher. Please contact Rutgers University Press, 106 Somerset Street, New Brunswick, NJ 08901. The only exception to this prohibition is "fair use" as defined by U.S. copyright law.

References to internet websites (URLs) were accurate at the time of writing. Neither the author nor Rutgers University Press is responsible for URLs that may have expired or changed since the manuscript was prepared.

⊖ The paper used in this publication meets the requirements of the American National Standard for Information Sciences—Permanence of Paper for Printed Library Materials, ANSI Z39.48-1992.

www.rutgersuniversitypress.org

Manufactured in the United States of America

A ship in harbor is safe—but that is not what ships are built for.

—**JOHN A. SHEDD**

Contents

	Foreword by Michael Aaron Rockland	ix
	Preface	xiii
	Introduction	1
1	Early Historical Background	8
2	The Post–World War II Era	25
3	The Invention of Containerization	37
4	The Rapid Growth of Containerization	53
5	From the Ocean to the Docks	70
6	Navigation	92
7	Pilotage	115
8	Tugboats	135
9	The Contemporary Port	151
10	Moving the Freight	172
11	The Seamen's Church Institute	189
12	The Future	211

Acknowledgments	227
Notes	233
Index	253

Foreword

It is with pride and pleasure that I pen this foreword introducing Angus Kress Gillespie's fine new book on Port Newark and the origins of containerization. The latter has radically altered the shipping industry and trade around the world, and its origins were in Port Newark and more recently in Elizabeth as well. When one considers how long New York fought against New Jersey having ports of its own, insisting until 1834, when the U.S. Supreme Court decided otherwise, that its western border was the Jersey shore, not the middle of the Hudson River, it is ironic that today the major segments of the port functions of the Port Authority of New York and New Jersey, a bistate agency, are in New Jersey.

It is not accidental that this book of Gillespie's has a nautical bent. He has long been a member of the Navy League and regularly invites Navy and Coast Guard officials to his classes at Rutgers University. The sea is very much in his DNA. His father and brother both had careers as doctors in the Navy. This was similarly true of his grandfather, and looking even further back, his great-grandfather, while not in the Navy, was commissioned by Abraham Lincoln himself to serve in the Union Army. Gillespie proudly has the commissions of all of his family framed and mounted on the walls of his home.

I have long been not only familiar with but a fan of Gillespie's work. This has especially been true of his four most recent books, including the present one. They share the following characteristic: an examination of a large and essential institution—how it was conceived, how it was built, and how it works. The second book, *Twin Towers* (Rutgers University Press), was written shortly before the terrible tragedy of 9/11/2001 and remains the indispensable resource for understanding the nature of the World Trade Center and its unique characteristics, available to historians and general readers now and into the distant future. The third book, *Crossing under the Hudson: The Story of the Holland and Lincoln Tunnels* (Rutgers University Press), again examines vital structures that link and animate our part of the world.

I've been rather reticent of mentioning the first of these four books because I was honored to be Gillespie's coauthor, and coauthors do not normally write forewords to books. Let me say at once that the idea was entirely his. Indeed, when he approached me, I anticipated little interest in a book about a road. Further, friends and colleagues warned me about getting involved in the project. How wrong they (and I!) would prove to be. The New Jersey Turnpike is not just a road; it is *the* road, probably the most important and most heavily traveled road in the world. *Looking for America on the New Jersey Turnpike* (Rutgers University Press, 1989), its title derived from the last lines in Paul Simon's song "America," uses this road, as we did, to tell America's story. The book has turned into something of a classic. It remains in print and continues to sell, and a new edition is in the works. Both Gillespie and I are still invited to give talks on a book that, as I write this, appeared thirty-one years ago. The *New York Times* back then gave the book a splendid review, and its authors were frankly shocked when the New Jersey State Library issued a list of "Ten Best Books Ever Written on New Jersey or by New Jerseyans," and we found ourselves accompanying the likes of Walt Whitman, Philip Roth, Allen Ginsberg, F. Scott Fitzgerald, and Stephen Crane on the list. I shall always be indebted to Gillespie for including me in a project that went on to achieve such distinction.

He has a genius for coming up with books about indispensable institutions and structures that we pass through or frequent without being conscious of them. And now he's done it again. As with the Turnpike book, the World Trade Center book, and the tunnels book, here he uses New Jersey's port and the ingenious new way of shipping developed there to tell the story of a people, our people. Gillespie has a particular taste for the vernacular, always finding meaning in the ordinary. This, perhaps, is at least in part because he was trained as a folklorist. With that background, it is natural with him to find human meaning in inanimate structures. It is the study of material culture on a massive scale.

But it is something more too. Gillespie has an uncanny ability to invest his subjects with affection and even love. He deserves our appreciation for awakening in us, his readers, those very sentiments.

Michael Aaron Rockland
Professor of American Studies Emeritus
Rutgers University

Preface

Here I would like to briefly explain the origins of this book, a story of how I came to write it. Coming from a long line of seafarers, and living in New Jersey, it was perhaps inevitable that I would become fascinated with the ports at Elizabeth and Newark. Driving north on the New Jersey Turnpike, I would often look over my right shoulder and see the giant cranes and the stacks of multicolored containers. I drove past the ports many times, and I kept saying to myself that I really should take a closer look. Driven by curiosity, I finally set aside a day to see these two ports for myself. But where to begin? After poking around on the internet, I decided to visit the center for the Seamen's Church Institute (SCI) at 118 Export Street in Port Newark. In retrospect, that was, for sure, the best place to begin.

Some years ago, I set aside a day to explore the ports. I drove up the New Jersey Turnpike, took Exit 14, and followed the signs for Port Newark. It's a place where few people visit unless they have some business there. I must confess that my first visit there was very intimidating. Driving in the port in my little car, I was dodging big tractor-trailers hauling multicolored containers with names like Maersk, Hanjin, Evergreen, and Overseas Orient Container Lines. Arriving at SCI, I introduced myself to the director, the Reverend Jean R. Smith, and I explained to her my interest in the ports. She was most cordial and hospitable. She showed me around the center and explained what her agency does for seafarers and port workers—including truckers, stevedores, and warehouse workers.

Smith explained that modern seafarers often suffer from feelings of isolation and loneliness aboard ship. The hours are long, and there is little social cohesion onboard. The problem is compounded by the fact that the ships have quick turnarounds in port. So the work of the SCI in helping the seafarers during their brief time in port is crucial. There are practical services. The center offers them free Wi-Fi and the use of computers. Even more important, they offer international telephone service and low-cost phone cards, so that the seafarers can call home. In addition, they provide an honest, low-cost money transfer service. Equally important, the ministers and the staff offer hospitality—a friendly face and someone to listen. Almost at once, I felt a deep appreciation for her work, and I scheduled a number of repeat visits.

In time, Reverend Smith encouraged me to accompany chaplains on visits to the ships in port. It was a wonderful opportunity, but there was some preliminary paperwork. It turns out that you cannot just drive into the secure area where the ships are docked. You first need to apply for a Transport Workers Identification Credential (TWIC). It has an ID card encased in heavy plastic that you must wear around your neck to get access to restricted areas of maritime facilities. The U.S. Transportation Security Administration (TSA) and the U.S. Coast Guard administer the TWIC. So I went to the TWIC office and filled out the paperwork. In a few weeks, I got my TWIC card. Now I was all set.

These ship visits were real eye-opening experiences for me. The chaplains would climb the long gangways up the side of the ship and be welcomed aboard, typically shown to the ship's galley, where they could meet with the ordinary seafarers. There were practical matters. The chaplains could offer phones, phone calls, SIM cards, and "top ups." The chaplains could also offer money transfer services. Most importantly, the chaplains were good listeners. They could help with problems, and find solutions. And, sometimes, the best thing to do was to pray together with the seafarer.

When the seafarers arrived in port, oftentimes the SCI would arrange a free shuttle service from the terminal gates to the nearby Jersey Gardens Outlet Mall with some two hundred stores, all under one roof. Typically, I would accompany the minister doing the driving there and back. On occasion, when they were shorthanded, they would let me drive the van. I remember vividly one day, while driving over to the mall, there was a young Filipino seafarer, in the back of the van, talking on his cell phone with his wife back in the Philippines. She was giving him her lengthy shopping list. I could not help but see the irony. He was about to shop for stuff that quite possibly he and his shipmates had brought over to the United States from China on a previous visit.

Over time, I found the port less intimidating and overwhelming. It was still exciting and interesting, but it gradually became familiar. It became clear that most all of the containerized cargo was arriving at the Elizabeth side of the port, while the Newark side was receiving all kinds of specialized bulk cargo. For example, on the Newark side, we find the High Bridge Stone Company, a wholesale supplier of Belgian block. Of course, today these hand-cut, granite cobblestones are coming not from Belgium but from India. You can't go there to pick up a few pieces to landscape your backyard, as everything in their vast inventory is sold by the ton. Also on the Newark side, we find LaFarge Gypsum, a company that imports tons of gypsum, the raw ingredient used to make wallboard throughout a network of plants in United States and Canada.

In all of these early explorations, I benefitted from my affiliation with the SCI clergy and staff members, who would explain the workings of the port to me. For example, I spotted a huge tank farm at Port Newark. What were these tanks for? It turned out that they were owned by Cutrale Citrus Products, the world's largest supplier of orange juice, with a 60 percent share of the market. In the newest section, there were four 650,000-gallon refrigerated storage units. I learned that more than one-third of the orange juice concentrate that enters the United States moves through Port Newark and 96 percent of that comes from Brazil. When you go to the supermarket, most of the orange juice that you buy is a blend of domestic and imported juice. Because Brazilian juice is tarter, it is often cut with the sweeter, domestic juice. Interestingly enough, New Jersey is central to orange juice processing and distribution in the United States because the imports arrive at Port Newark.

As I look back on those early visits to the port, I realize that I came away full of curiosity, wanting to learn more. It was there that I first got the idea for this book, which has taken far longer than I first imagined. Why did it take so long? The story has many moving parts—the ships, the terminals, the trucks, the railroads, and the warehouses. Not to mention the investors, the managers, and the workers. There is an overwhelming amount of material out there. I have tried to grapple with this material, to shape it so as to make it a readable account. I hope that you will allow me to be your guide as we explore Port Newark–Elizabeth together.

Port Newark and the Origins of Container Shipping

Introduction

• • • • • • • • • • • • •

How best to tell the story of Port Newark? I decided early on that there were really two stories. The first is a narrative history of the port from the early days of the 1600s right up through the 1950s. This story, largely based on archival research, takes up the first four chapters, about a third of the book. These chapters trace the gradual evolution of Port Newark from a sleepy colonial seaport to its current status as a major container port. Two things of great significance happened in the late 1950s. First, there was the invention of containerization. Second, almost simultaneously, there was the expansion of the port southward into the immediately adjacent city of Elizabeth. This expansion was explicitly designed to handle containers. Thus, technically we now had two ports—the original Port Newark and the brand-new Elizabeth Marine Terminal. In practice, nearly everyone thinks of these two as a single port. Thus, in this book, I refer to it as simply Port Newark–Elizabeth in the later chapters.

The second story is an attempt to explain the workings of the port at the present time. Based largely on interviews with key players, it constitutes the final eight chapters of the book. Beginning in 2012, I spent several years at the port asking lots of questions. I was full of curiosity about how things work. There is a large and fascinating cast of characters who do the work of the port. There are the mariners who steer the big containerships. They are assisted by both the bar pilots and the docking pilots as well as the tugboat operators. There are the Coast Guard personnel responsible for vessel traffic control and aids to navigation. There are the longshoremen who load and unload the ships. The list goes on and on. Not everyone was willing to speak with me. Some appeared to fear that I was going to write a muckraking book, a sort of exposé. Others may have simply been too busy to bother. In any event, I am very grateful for those who took the time to speak with me, and they have their rightful place in my acknowledgments.

From a maritime point of view, as we look at a map of the U.S. East Coast, there are many interesting ports, including those in Boston, Philadelphia, and Baltimore. Each has its own story worth exploring, but what's special about Port Newark? Well, for one thing, it is a New Jersey story. I am pleased and proud that this book is being published by Rutgers University Press, which has a focus on New Jersey and the surrounding region. Without a doubt, all New Jerseyans can take pride in Port Newark as a monumental economic engine for the state. At the same time, it has not just statewide significance but also *worldwide significance*. Why? Because it is the birthplace of containerization.

Containerization started at Port Newark on April 26, 1956. Some fifty-eight containers, each thirty-three feet long, normally transported by truck, were instead loaded aboard the SS *Ideal X*, a converted oil tanker, headed for Houston, Texas, where it arrived six days later. It was a revolutionary development. They did not put the whole truck trailer on the ship. The trailers were removed from their steel beds, axles, and wheels. All that was left was the trailer bodies, and these could be stacked. The inventor used one ship in place of many trucks, with big savings. His name was Malcom McLean, and his company was called Sea-Land, though today it is called SeaLand.

In this book, I argue that his development of containerization should rank right up there with Henry Ford's development of affordable automobiles. Of course, I realize that this is a tough argument. I noticed a problem as I was doing research for this book. As I talked over my project with family, most colleagues, and students, I came to realize that few had ever heard the name Malcom McLean. At the same time, all my friends in the maritime industry, without exception, knew the story quite well. I find the story both interesting and important, and I hope to make it better known.

In chapter 1, "Early Historical Background," I set the stage for McLean's launching of containerization. In the nineteenth century, Newark became a prosperous manufacturing city with a rudimentary shallow port on the Passaic River. Then, in the early twentieth century, the City of Newark began digging a deep-water channel through the swampy meadowlands of Newark Bay. Of course, there was considerable expansion during World War I and continuing through the twenties. After the setback of the Great Depression, there was another expansion during World War II. However, during this entire period, the ports on the New York side of the harbor dominated trade in the region.

We continue the historical narrative with chapter 2, "The Post–World War II Era." At this time, Port Newark was owned and managed by the City of Newark, but it was struggling and seriously undercapitalized. After lengthy negotiations, the Port Authority agreed to take over its management in 1947. As part of the agreement, the Port Authority promised to modernize and improve the facility. In addition, they agreed to pay a substantial sum to the city every year in lieu of taxes. The Port Authority was just what was needed to bring Port Newark up to date. The other big development during this period was an agreement in 1955 with the State of New Jersey for the Port Authority to expand the port into the city of Elizabeth, another huge success.

In chapter 3, "The Invention of Containerization," we take up that significant moment in history—the first commercially successful shipment of containers from one port to another. Today, of course, we take container shipping for granted. It is part of our everyday landscape; but, in 1956, it was a revolutionary breakthrough. In this chapter I explain how one man—Malcom McLean—was able to think though the concept, overcome the obstacles, and begin a revolution in shipping.

Of course, McLean did not accomplish all this by himself. He assembled a talented team of experts. There were many of them, but we should mention at least three. First, there was Keith Tantlinger, a technical genius, who took care of details like designing the corner fittings of the containers so that they could be engaged with other containers. Then there was naval architect Charles R. Cushing, who helped to adapt older ships for containerization, thus sparing McLean the expense of buying new ships. Finally, there was Walter Wriston of the National City Bank of New York (later Citibank), who saw the potential in investing in McLean's vision.

In chapter 4, "The Rapid Growth of Containerization," I trace McLean's efforts to expand his business by opening up more ports dedicated to containerization. He turned to a young civil engineer on his staff, Ron Katims, to make that happen. In the 1960s, McLean focused on ports under the American flag, protected from foreign competition by the Jones Act of 1920 that specified that commerce between any two American ports had to be transported by ships that were built, owned, and operated by U.S. citizens. How to proceed? Each case was different, but the expansion followed a three-step pattern. First, was the prospective port authority interested in the idea? Second, there was an extended negotiation. Who would pay for the construction? What were the terms of the lease? The third step was straightforward—the building of the port. Katims was able to build ports in Puerto Rico as well as in both Oakland and Long Beach in California. Certainly, most challenging was the effort to persuade the U.S. Army to accept containerization for sending supplies to Vietnam. Katims had to use all of his powers of persuasion to get skeptical Army engineers and port operators to accept containerization. In the end, it worked out. The rest of the world was watching, and the concept was quickly catching on.

Chapter 5, "From the Ocean to the Docks," represents a transition from the historical section of the book to the more contemporary section. Here we take an imaginary ride onboard a large containership as it passes from the ocean, through the harbor, and into the terminal. We, as riders, have the opportunity to take note of the rich history, on both sides of the channel, as we ride along. We take note of the many landmarks including historic lighthouses, forts, and a number of small islands. As we go under the Verrazzano-Narrows Bridge, we pass from Lower New York Bay to Upper New York Bay. Later, we make a sharp left turn into the narrow Kill van Kull, and soon we pass underneath the Bayonne Bridge. Then we soon come upon Bergen Point on the north side of the Kill van Kull, and we take a hard right turn into Newark Bay. We now have less than a mile to our destination, Port Newark–Elizabeth.

As we are taking this imaginary sightseeing journey, we are enjoying the scenery and probably not thinking very much about the crew on the bridge responsible for steering the ship through the busy harbor. However, now in chapter 6, "Navigation," we turn our attention to that problem. It turns out that the crew on the bridge gets a great deal of help from the U.S. Coast Guard, which maintains a large number of "aids to navigation" (ATON) that include buoys of many shapes and colors used to mark the sides of channels. They also maintain fixed aids, or daymarks, which are often on pilings on the water. This task of maintaining navigation aids on the water can be compared to the work of a state highway commission on land. In addition the Coast Guard has an office of Vessel Traffic Service (VTS), which has often been compared to Air Traffic Control. The people in this office monitor the ships in the harbor using radar, closed-circuit television, VHF radiotelephones, and an automatic identification system (AIS) to provide navigational safety.

To be sure, the U.S. Coast Guard provides a good deal of help in getting ships in and out of the harbor, but the story does not end there. In chapter 7, "Pilotage," we consider the local navigational experts who do the actual work of ship steering in narrow and congested waters. There are two basic types of maritime pilots serving in the New York–New Jersey Harbor, with two different sets of skills. The first group are known as "bar" pilots. They are the ones who bring ships in from the open ocean into the Lower New York Bay and past the Verrazzano-Narrows Bridge. After that, at some point, the bar pilot hands off control to a "docking" pilot. The docking pilot, with the aid of tugboats, takes the ship through the channels and nudges it into a berth.

In chapter 8, "Tugboats," I follow up on the docking procedures. Of course the docking pilots are in charge of the evolution, but they are highly dependent on the work of the supporting tugboats. These are small, powerful vessels that assist in docking huge ships that lack maneuverability on their own. They work in teams under the direction of the docking pilot. Some of the tugs maneuver the big ship by pushing on clearly marked reinforced sweet spots. Others can pull the big ship by ropes attached to bollards, short and sturdy posts, on the ship. Most tugboats used for docking larger ships today are equipped with what is called a Z-drive, a special kind of marine propulsion. These tugboats typically have twin screws, each of which can rotate 360 degrees. This ability permits rapid changes in thrust direction. Understandably, these tugboats are highly maneuverable, nimble, and efficient.

During the period of time that I was working on this book, I had the sense of sometimes trying to hit a moving target. There were many changes taking place, so I report on them in chapter 9, "The Contemporary Port." Most of the major changes were made in an effort to cope with the trend in the maritime industry toward bigger container ships with deeper drafts. This part of the story begins with the start of work in 2004 with the goal of providing fifty-foot depths in the channel all the way from the Atlantic Ocean to the Newark Bay Channel. It was a wise decision because, without a deeper channel, the newer and larger containerships would end up simply going elsewhere. Pressure to adapt to change kept increasing. As the dredging of the channel was proceeding, the Republic of Panama announced the expansion of the Panama Canal in 2006. It now became clear it would be necessary to raise the roadway of the Bayonne Bridge in order to allow larger containerships to access the port. Raising the bridge was expensive, difficult, and controversial, but it all somehow worked out. It was on September 7, 2017, that the first Panamax ship, the *T. Roosevelt*, passed underneath the new Bayonne Bridge and entered Port Newark–Elizabeth.

On the Jersey side of the larger Port of New York and New Jersey there are five container terminals. Though each one has its own way of doing things, they do have a great deal in common. For my purposes, I decided to focus on the APM Terminal at Port Elizabeth, the subject of chapter 10, "Moving the Freight." Interestingly enough, this is the same terminal that was described in chapter 4, the first terminal that was purposely built for containerization by McLean for his company Sea-Land. Over the years, this terminal has been greatly upgraded, and it has changed corporate hands from Sea-Land to Maersk. Yet to the alert observer, it is still recognizable. In this chapter, I explain in some detail how the cranes get the containers on and off the ships as well as how the trucks bring some of the containers to the port for export and how they bring others out of the port for delivery. Finally, I explain how port managers keep track of all these comings and goings.

Obviously, Port Newark–Elizabeth is an American port. The offloaded containers are handled by American longshoremen, and these containers are taken to their destinations by American railroaders and American truckers. What is not so obvious is that the ships that bring those containers from all over the world to us are all sailing under foreign flags, such as Panama, Liberia, Belize, or the Marshall Islands. These ships are staffed by hardworking crews from places like the Philippines, Indonesia, Ukraine, China, Pakistan, and Sri Lanka.

It is a hard life. Seafarers spend many months at sea away from friends, loved ones, and family. So who looks after the needs of these mariners? Fortunately, there are a number of agencies dedicated to this task. The largest of them is the Seamen's Church Institute (SCI, the subject of chapter 11).

The chaplains of SCI help out in a number of ways. The first and most obvious is when a chaplain visits a ship in port. The chaplain can help in practical ways such as providing low-cost SIM cards for cell phones and helping with the transfer of money to their families back home. Chaplains refrain from proselytizing, but they can assist crew members with prayers or Holy Communion upon request. Oftentimes, the seafarers have little time in port, so a ship visit must suffice. For those with more time, they may be able to visit the International Seafarers' Center located within Port Newark. There the seafarers are welcomed with hospitality, and they are able to access high-speed Internet and many other amenities including chapel, lounge, and workspace areas.

What will the future of ships and ports be like? I deal with this question in chapter 12, "The Future," where I look first at the future of ports and then later at the future of ships. With regard to ports, there are risks ahead such as storms, floods, power outages, pandemics, trade wars, and cyberattacks. On a more positive note, there are real possibilities for port automation. There is the very real prospect of gains in operating efficiency, along with lower labor costs. These are exciting developments, but I believe that they will happen slowly in the United States because of strong longshore unions on both the West Coast and the East.

As we look ahead to the future of ships, the industry will have to cope with traditional risks such as bad weather, piracy, dangerous cargoes, and structural flaws. Just as we have seen with ports, ships may have a promising future with more and more automation. A review of the literature indicates that maritime autonomous surface ship technology is a hot topic. We consider whether it can happen. In terms of the technology, the answer is certainly yes. But will it happen? Here we begin to see a great deal of hesitation. There are really no strong maritime unions to oppose automation. Instead, the main problem to the widespread adoption of autonomous surface ships is the matter of law. Existing U.S. and international regulations were drafted without anticipation of automation. So we are faced with questions such as who is in charge and who is responsible.

Port Newark and Port Elizabeth together have certainly played an important role in shaping New Jersey's geography, history, and economics. However, if we dig a bit deeper, we also realize that this is the place where international trade in containerization got started. Given the United States has a consumer-based economy, it makes sense to spend some time with a book like this to better understand American consumer culture. We Americans are accustomed to having many choices whenever we go shopping, whether at the supermarket, the dollar store, or online. We have plenty of choices when looking for affordable personal computers, cosmetics, cell phones, televisions, home furnishings, apparel, or footwear. In this book I give the reader an easily understandable and nontechnical explanation of how it all came about.

1

Early Historical Background

• • • • • • • • • • • • • • •

In this chapter, we will go back in time to the earliest days on the New Jersey side of the harbor. We need this background information to fill in the crucial backstory—to understand the landscape that gave Malcom McLean the opportunity to form his successful shipping business. We are going back to witness firsthand those formative moments in Newark. This chapter takes us deeper into the hearts and minds of the men who laid the groundwork for McLean's success.

A group of Puritans from Connecticut first settled Newark, New Jersey, in 1666 along the banks of the Passaic River. The settlers were led by Robert Treat, who at age forty-four had established himself as a key leader in the New Haven Colony, with a proven record as an experienced land surveyor, tax collector, military officer, and judge. Treat purchased the land directly from the Hackensack Indians for a mixed group of trade goods including gunpowder, bars of lead, axes, coats, guns, pistols, swords, kettles, and blankets. Newark began as a small agricultural community, but over time became a leading manufacturing center due in large measure to its access to New York markets.

By the early 1830s, the population had grown to fifteen thousand with more than seventeen hundred buildings, most of them built of wood. Newark prospered as a seaport since vessels could easily transit from the Passaic River to

Newark Bay and on through the Kill van Kull to New York City. Inbound vessels brought building materials for the growing port town, and outbound vessels carried coal and iron—brought from inland by way of canal and rail. The steamboat *Newark* made twice-daily trips to New York, with some seventy-five passengers each trip.[1]

In the years leading up to the Civil War, Newark became a prosperous manufacturing city. The Passaic River was busy with paddle wheelers and schooners bringing in raw materials and taking out all kinds of finished goods. By 1846 Newark had more than a hundred factories; most of them driven by up-to-date steam engines rather than by old-fashioned water wheels. Factories manned by immigrants from Ireland and Germany turned out lace and saddles and jewelry and soap. Peter Ballantine brewed oceans of light lager beer. Some factories made steam engines; others ground cattle bones into fertilizer. The list of products seemed endless. It included rubber manufacturing for waterproof garments as well as America's first white zinc oxide for paint.[2]

In the aftermath of the Civil War, many new heavy industries emerged. The old prewar items such as beer and jewelry and paint were still being made. But now the city was turning out chemicals, hardware, cotton thread, enameled goods, and dozens of other products. The city leaders held a showcase of Newark goods at the great Industrial Exhibition of 1872. Visitors could see a complete array of Newark products. Historian John T. Cunningham explained, "They saw harnesses valued at $10,000, gold-plated sleigh bells worth up to $200 a set, one hundred styles of table oilcloths, books printed in Newark, pearl buttons, ribbons and a hundred varieties of paint. They saw buggies, walnut bird cages, steam fire engines, gold pens, chalk, tools, toys, malleable iron castings, telegraph instruments, and clocks."[3]

Things were going great for Newark; but, by the turn of the century, the city's lack of a deep-water port was putting the brakes on further development. For its first two hundred years, the city made do with the docks along the right bank of the Passaic River. These docks were along the high land of Newark, where the first settlers touched shore, four miles inland from Newark Bay. It was an entirely satisfactory arrangement for the shallow-draft paddle wheelers and schooners of the eighteenth and nineteenth centuries. But now the city fathers had their eyes on the wide waters of Newark Bay. But there was a serious problem. The land where Newark sloped off to the bay was made up of marshy meadowlands. How to solve the problem? Dig a deep ship channel and build up some huge docking areas. They needed to dig a navigable channel where there had been only shallow water and to create solid land where they had been none. Only then could they take full advantage of all the economic opportunities that Newark Bay had to offer. It was a three-step process.[4]

Of course, in those early days, the wetlands were seen as totally without value. They were "swamps" or "bog" in need of being transformed into productive land. There was no need to file an environmental impact statement, as would be required today. We now know that wetlands are valuable systems in the natural scheme of things. The World Wildlife Fund's *Atlas of the Environment* explains, "They provide critical habitats for thousands of species of plants and animals, yield up food, fiber, and building materials, play important roles in regulating water cycles, filter pollution and guard shorelines from the depredations of the sea."[5] Port Newark was a product of its time. It was a totally artificial landscape, with any threat from nature subdued. Because of environmental regulations, I think it is safe to say that such a port could not be built today.

In any event, as a first step, the city began, without fanfare, to quietly buy up meadowland property off Newark Bay starting in the early twentieth century. Officials were careful to be discreet in order to avoid stirring up interest among land speculators. Understandably, there were legal problems in acquiring clear titles because the land had long been considered worthless and forgotten, home to muskrats and mosquitoes. Some of the titles went back to the seventeenth century, and there was no clear ownership. By 1911 the city had spent nearly one million dollars buying some 930 acres of meadowlands property along the bay. So the city acquired a port zone that had a frontage of 4,000 feet along Newark Bay, going back inland some 13,000 feet.

As a second step, in 1912 the city authorized spending another million dollars to dig a channel, some 300 feet wide and 20 feet deep through the swampy meadowlands from Newark Bay into the city. Actual dredging began on New Year's Day 1914. The mud was sucked from the bottom of the channel by means of a large hydraulic dredge; the mud was then forced through a 27-inch-diameter pipe into a space confined by dikes, then drained slowly, reclaiming 156 acres of marshland. This filling process raised up the level of the reclaimed land a full six feet above the surrounding muck.[6]

As a third step, the city began building the main dock or landing pier. It ran a full 4,000 feet along the northerly edge of the channel. It included a track for a traveling derrick, a large crane used for lifting heavy cargo in the process of either loading or unloading the vessels. There was also a freight track of standard gauge, to connect with the nearby rail lines—Jersey Central, Pennsylvania, and Leigh Valley. Administrators set aside large lots to be leased to manufacturing plants, with ready access to both rail and sea. Finally, they put in wide streets across the meadows to connect the city with the reclaimed areas.[7]

By June 1915, chief engineer Morris R. Sherrerd said, with understandable pride, "The manufacturers located here will have the best possible highway to and from the large markets of the world. We will offer the best factory sites bordering on the best water-rail terminal on the Atlantic Coast. We have provided an ideal industrial centre, combining the best service in ocean and industrial traffic. In other words, we are going to add value considerably to the manufacturing prestige of the United States of America, and we need the safety secured only through greater export trade. Newark is laying the foundation for this safety and offers to the nation a practical port."[8]

Notwithstanding Sherrerd's grandiose statement, the simple fact was that the largest seaport on the East Coast at that time was in Brooklyn and at the finger piers on the West Side of Manhattan. As we shall see, it would take many twists of fate and many years for Port Newark to become dominant.

The First World War

World War I, a global war centered in Europe, began on July 28, 1914. For some two and a half years, President Woodrow Wilson tried to keep the United States neutral, but America finally entered the war in April 1917. With the mobilization for war came a tremendous expansion of Port Newark. There were two components to that expansion—the creation of a U.S. Army supply base and the development of a large shipbuilding installation.

Let us begin with the story of the U.S. Army supply base. A site of some 133 acres at Port Newark was purchased by the Army, at the direction of President Woodrow Wilson, from the City of Newark on November 15, 1917.[9] During the war, the main Army supply base was located in Brooklyn, New York. Stored there were millions of pounds of bacon, sugar, canned fruits, fish, meats, vegetables, tea, coffee, prunes, beans, and other foodstuffs. In addition, there were millions of uniforms, overcoats, hats, shoes, yards of khaki cloth, and muslin, not to mention small arms and medical equipment. The base at Brooklyn was the principal distribution center, but that base simply could not handle the tremendous volume of goods. To relieve the congestion, a supplementary port had to be located to store some of the goods in reserve, to be drawn upon as needed. Thus, great quantities of war materiel were diverted to Port Newark. As those supplies were required, they could be loaded aboard lighters, or unpowered and flat-bottomed barges, and towed over to Brooklyn to be placed onboard the ships on an "as needed" basis. In addition, a great deal of ordnance, including weapons and ammunition, was stored at Port Newark. Rather than haul such hazardous loads over to Brooklyn, workers put the ordnance on barges and loaded up the ammunition ships lower down in the bay, away from heavily populated areas.

The 133-acre Army reservation at Port Newark could deal with an enormous amount of material. The total for goods handled in and out of storage for the fiscal year ending June 30, 1919, was 419,350 tons. That tonnage was not as great as that handled at Kearny, New Jersey, but it involved a good deal of work. One of the great strengths of the Army facility at Port Newark was a bulkhead of some 3,600 feet in length, which could accommodate many lighters at once. That bulkhead was 150 feet wide, half of which was covered. On the uncovered portion of the bulkhead was a double railroad track. The inboard track could be used for the storage of rail cars. The outboard track held ten very large, movable locomotive cranes with hydraulic systems and controls. These cranes could load the lighters with heavy guns, slabs of steel, caissons, water carts, and rolling kitchens. The cranes, being movable, could be placed into position for loading and unloading, as needed. There were fourteen acres of open storage, and nearly every inch was piled high with ordnance.

The Army facility at Port Newark had fourteen miles of railroad tracks, enough to store some 700 rail cars. The warehouses were built at right angles to the bulkhead. These were one-story buildings about 1,100 feet long and 50 feet wide. There were nine of these buildings, separated by railroad sidings. Each building had a wide shipping platform. There were two large sheds for the storage of hay, oats, and automobiles—encompassing 600,000 square feet. In terms of trucking, the facility was equipped with 63 tractors and 720 trailers. An average day's work involved handling between 5,000 and 6,000 tons. During the war, this was a busy place; but after the war, the army depot became inactive.[10]

Alongside the U.S. Army base in Port Newark was a massive wartime shipbuilding yard. The shipbuilder was named Submarine Boat Corporation (SBC); but, oddly enough, they were not building submarines here. The SBC had extensive previous experience in building submarines, but they decided that it would be more profitable to turn to building cargo vessels instead of submarines. So they leased a large shipyard at Port Newark where they built twenty-eight slipways, structures consisting of paved slopes down to the water from the place where the ships were to be built. Thus they could work on twenty-eight ships at a time. They also outfitted piers and shops for the fabrication of some of the steel required.[11]

The SBC contracted with the Emergency Fleet Corporation to build 150 five-thousand-ton ships. During the First World War, the SBC's yard was the second largest shipyard in the United States, with some 25,000 men working there. To build the ships as fast as possible, the ship parts were made in steel shops throughout the country and then assembled at the shipyard. Most of the work in forming the parts was done in faraway shops, accustomed to building parts for bridges and office buildings, from drawings furnished by the SBC. The drawings were very precise, and the work had to be done carefully so that when the parts were brought together, they would fit together perfectly. The different parts were brought to the yard in sections as large as could be transported by railroad. Each part was numbered and lettered. The parts were placed into position and bolted. Finally they were fastened with rivets. Building the hull of a given vessel required the driving of more than 400,000 rivets.[12]

The first such standardized steel cargo ship to be built was the SS *Agawam*; the keel was laid down on December 20, 1917. The keel, of course, was the principal structural member of the ship, running lengthwise along the center line from bow to stern, to which the frames were to be attached. Laying down the keel was the formal recognition of the ship's construction, marked by a ceremony attended by dignitaries from the shipbuilding company and the government. In just a little over five months, the ship was launched and christened on May 30, 1918. The oceangoing steamship was completed and delivered in late October 1918, thus under construction for only 303 days, a remarkable record. Some 27 steel mills, 56 fabricating plants, 200 foundries, and machine, pipe, joinery, and equipment shops were involved. The *Agawam* had a displacement of about 7,800 tons when loaded and a deadweight carrying capacity of about 5,500 tons. It measured 343 feet long and 46 feet wide, with a depth of 23 feet. When fully loaded, it could maintain a speed of ten and a half knots at sea.[13]

The machinery for the vessel consisted of a Westinghouse steam turbine operating at 3,600 revolutions per minute, driving a single screw propeller at 90 revolutions per minute through a reduction gear. Steam supply to the turbine was furnished by two Babcock and Wilcox boilers. The auxiliary machinery of the vessel included electric lighting, a refrigeration plant, pumping machinery, and forced-draft blowers. Also found were heaters, cargo handling machinery, a steering engine, as well as the windlass, used for hoisting the anchor. Fuel oil was burned in the boilers for the generation of steam.

The fuel was carried in sufficient quantity to more than cover a round trip to Europe. Ample and comfortable quarters were provided for both officers and crew.[14]

Mrs. Woodrow Wilson was invited to select names for the Emergency Fleet, and she decided that they should all be of American Indian origin. For the first ship, she chose the name Agawam—meaning "Great Salt Meadows of the Atlantic Coast," certainly appropriate for a ship assembled at Port Newark. The sponsor at the launching on Memorial Day of 1918 was Miss Mary Eurana Ward.[15]

Five months later, on October 26, 1918, the *Agawam* was completed by the SBC and delivered to the U.S. Shipping Board's Emergency Fleet Corporation. The oceangoing steamship was completely finished and ready for sea. Edward N. Hurley, chairman of the Shipping Board, sent a congratulatory message to President Henry R. Sutpen of the SBC: "Please accept the congratulations of the successful completion of the *Agawam*, the first of our fabricated ships. This is an historical event in American shipbuilding. It marks the era actually of ship manufacturing and is, I believe, one of the great turning points in ship construction. Most assuredly, the employees and officials of the Submarine Boat Corporation may feel that, in building the first fabricated ship, they played roles in shipbuilding—in wartime shipbuilding—which they may always remember with patriotic pride."[16]

Conflict with New York State

By 1917, politicians in New Jersey had a distinct feeling that the ports on the New Jersey side were at an unfair competitive advantage with the ports on the New York side. There was a bitter conflict over the way railroads were pricing the handling of freight. To understand the conflict, we have to explain the "lighterage" problem. Let us imagine a shipment of goods coming into New York from the western United States. Let us further imagine that the shipment was coming by way of the Pennsylvania Railroad or the Baltimore and Ohio Railroad or the Lehigh Railroad. All of these lines had terminals on the New Jersey side of the Hudson. How to get the goods over to Manhattan? It was done by ferryboats in one of two ways. You could take the freight out of the rail cars on the New Jersey side, box by box, piece by piece, and put it onto barges and tow the barges across the river. This process was called "lighterage." Or you could just put the whole rail car onto a flat barge equipped with rails and tow it across the river. This process was called "car floatage." Either way, it was very expensive.[17]

In brief, the rail terminals were on the New Jersey side, but most of the goods were destined for the New York side. Both states benefitted from the economic activity of the port, but New York was the dominant partner since it had the larger market and was the center of international commerce and banking. Historically, the two states had gone through cycles of conflict and cooperation. But in 1917, the stage was set for conflict. The problem, as seen from the New Jersey side, was the way that railroad rates were set. For long-haul traffic, rates were based on a system of zones, rather than on exact cost or distance. Within a given zone, the rate was the same, regardless of cost or distance. Let us use an example from the New York subway system. Imagine that you are taking the number 1 train, or the Seventh Avenue Local, northbound, from South Ferry, or Battery Park, in Lower Manhattan. You pay the same fare whether you are going all the way to Van Cortlandt Park in the Bronx or get off halfway at Times Square.[18]

New Jersey officials were understandably unhappy with the existing rate system. Since goods destined for New Jersey did not require expensive lighterage, why should shippers have to pay the same rate as goods sent across the Hudson River? Obviously the actual cost of shipping goods to New York was far greater than the cost of shipping goods to New Jersey. If the rates to New Jersey were lowered, the state would enjoy a competitive advantage. Major shipping lines might be attracted to Jersey City, Hoboken, and Newark. There would be new industries and more jobs. So a suit was brought before the Interstate Commerce Commission (ICC) by J. Spencer Smith, head of the New Jersey Board of Commerce and Navigation.[19] The situation was explained succinctly by Jameson W. Doig: "In treating the New York–New Jersey region as a single zone in setting freight rates, therefore, the railroads actually neutralized the low-cost advantage of the New Jersey side of the harbor."[20]

New Jersey had issued a challenge, and the New York Chamber of Commerce lost no time in responding. Eugenius Harvey Outerbridge, chairman of the Committee on Harbor and Shipping of the Chamber, was alarmed. So was Irving T. Bush, the creator of the Bush Terminal Company. The two men joined forces and assembled a legal team, headed by Julius Henry Cohen, to fight off this challenge. The hearing was held in the wood-paneled Great Hall of the Chamber. In the front of the hall hung the imposing and stern oil portrait of Cadwallader David Colden, the nineteenth-century district attorney for the First District, made up of Suffolk, Queens, Kings, Richmond, and Westchester counties. Under that portrait sat Wilbur LaRoe Jr., the chief examiner of the ICC. In front of him were placed, in a semicircular arrangement, three large tables—one for the New Jersey attorneys, one for the railroad attorneys, and one for the New York attorneys. The ensuing debates were polite and civil, but the stakes were very high.[21]

After listening to arguments from all sides, the ICC handed down its decision on December 17, 1917. The ICC sided with New York, stating that historically New York City and northern New Jersey "constitute a single community" and that New Jersey had benefitted by its proximity to New York. It was a bitter disappointment for New Jersey. At the same time, the commission agreed that the regional freight system was in need of change, but it did not accept the argument that New Jersey should be given a lower rail rate. However, the commission went on to criticize the lighterage system as burdensome and inefficient, holding out hope that it might revisit the issue if the two states were unable to solve the problem themselves.[22]

Creation of the Port Authority

The Port Authority of New York was created in April 1921 by an agreement between the two states with a mission to improve transportation and terminal facilities in the Port District, which included all of New York City and much of northern New Jersey. The hope was that the new agency would be nonpolitical and have a passion for efficiency, honesty, and competence. To be sure, the creation of this new agency was a step in the right direction—a recognition that the two states both had a stake in the harbor. However, there was a problem from the very beginning. Consider that name, the Port Authority of New York. When we stop to think about it, that name clearly favored one state over the other. It was not until 1972, more than fifty years later, that the name was officially changed to the Port Authority of New York and New Jersey.

Historian Jameson Doig has pointed out that the Port Authority grew out of the faith in the rational planning of the Progressive Era. During the 1920s, the new organization had no interest in Port Newark. That would come later. What the Port Authority wanted to do in those early years was to construct rail tunnels that could connect Manhattan and Brooklyn to the mainland over in New Jersey. They would also build large freight buildings in Manhattan. Goods from all the railroads would be sorted, and the organized freight would be delivered by truck, increasing efficiency, saving money, and reducing traffic congestion in Manhattan. The cost of such tunnels and freight terminals would be justified only if all of the railroads agreed to end the tug and lighterage system for bringing freight into Manhattan. It was a rational plan backed by efficiency experts.[23]

To mobilize public opinion in support of the plan, the Port Authority created, in July 1921, a motion picture that dramatized the problem. It was animated cartoon featuring a potato with not only eyes but also a nose and ears, arms, and legs. We first see Mr. Potato in a freight train somewhere in New Jersey: "I've been in this train twenty-six hours," weeps Mr. Potato. "It is terrible."

Eventually, the train moves and reaches Jersey City. Mr. Potato gets out of the train and is dumped onto a barge, which takes him over to Manhattan. He is then placed on a truck, which moves slowly through dense traffic to the Washington Market. By this time, Mr. Potato is very tired indeed. He lies in the market for several hours, and he is then placed on a truck to retailer in the Bronx. We are told that it took Mr. Potato nine and a half hours to get from New Jersey to the retailer, at a cost of 68 cents. The cartoon next shows an alternative vision, doing away with the barges and bringing the railroads directly into Manhattan. Now the trip costs only 38 cents.[24]

The film made a persuasive and rational argument, but powerful forces were lined up against any reform of the freight-handling system. Tammany Hall was opposed to any cooperation with New Jersey or with the Port Authority. The railroad bosses were unwilling to cooperate with each other; they were wedded to a competitive model. The loading and unloading of freight cars and barges on either side of the Hudson was "extraordinarily expensive and laborious," but labor boss William J. McCormack, "Mr. Big," liked things just the way they were. In short, the railroads and New York City and Mr. Big all turned down the Port Authority's plan and thus doomed themselves. "So the piers rotted away and floated out to sea, along with the city's hopes for economic revitalization of its waterfront."[25] Thus in 1924 the Port Authority gave up and moved on, beginning work on a series of less controversial vehicular bridges connecting Staten Island and New Jersey—Bayonne, Goethals, and Outerbridge.[26]

The Roaring Twenties

For years, the New Jersey congressional delegation had tried to get federal funds for the expansion of Port Newark. Caught up in the expansionist enthusiasm of the Roaring Twenties, the City of Newark decided that it could wait no longer for federal backing. In 1921 the city issued $1.25 million in bonds to deepen the channel in Newark Bay from the Kill van Kull to the Port Newark Terminal. The work of creating a navigable 30-foot-wide channel basically involved three steps. First there was the actual hydraulic dredging. Second, the engineers created timber retaining bulkheads. Third, the dredging spoils were placed behind the bulkheads to create 300 acres of additional land south of the city channel. In addition, three large finger piers, 200 feet wide and 1,200 feet in length, were constructed, extending into Newark Bay. The new piers were provided with direct rail connections and access to connecting highways.[27]

The economic benefit of expanding Port Newark to the city was felt almost immediately. Throughout the 1920s, the city enjoyed unprecedented industrial growth and rising realty values. There was an influx of new industries and a boom in housing construction. Everything seemed to be feasible. Expressing the optimism of the era, Charles F. Kramer of the Newark Real Estate Board said, "Nature has endowed us with a strategic position. Those who selected this site as a city had great foresight.... Our seaport development is nearing the point where we can shelter the world's commerce. Our great transportation questions, both passenger and freight, are rapidly being planned and coordinated so that we will soon be the terminus of all these giant carriers."[28]

Despite Kramer's bold assertions, Port Newark was decidedly subordinate to the Port of New York throughout the 1920s. By every metric—total tonnage or total value of cargo—Port Newark lagged behind Manhattan and Brooklyn. Nonetheless, the City of Newark kept trying—and sometimes succeeding—to recruit new tenants for the port. For example, in early 1926, Newark successfully negotiated a twenty-year lease for a 30-acre site at Port Newark with the Weyerhaeuser Timber Company, one of the largest lumber companies in the nation. At that time, the company operated a fleet of steamers that made Port Newark its eastern terminal, where they hired some 300 workers. Over the period of the lease, it received 14 million feet of lumber monthly, with an annual value of about $10 million.[29]

Weyerhaeuser was in the business of bringing in lumber for the construction boom taking place in the New York–New Jersey region. The lumber was brought in from Nova Scotia as well as from the U.S. South and the West Coast. Besides lumber for framing construction, there were many bundles of lath, thin and narrow strips of wood used in the "lath and plaster" technique for the making of interior walls. The company also brought in large numbers of railroad ties."[30]

Late in 1926, there was another piece of good news for Port Newark. A tenant was found for the old army supply depot. The Mercur Trading Corporation leased the facility from the U.S. government for a period of years. It was still an attractive property with its waterfront facilities, warehouses, and miles of railroad trackage. Captain F. Jarka, chairman of the board of the company, said, "The Port Newark Terminal, with is advantageous railway connections and deep water, should prove attractive to shipping interests. The plant lends itself to the adoption of efficient operating methods, the development of which we are most concerned."[31]

Despite an occasional success, throughout the 1920s, Port Newark was clearly the underdog compared to the Port of New York. A study conducted by the Port Authority in 1929 attempted to account for the distribution of vessels docked in the region. A majority of them were found to be docked on the New York side, with only a much smaller number on the New Jersey side. According to the study, on a typical day there would be 19 vessels docked in Staten Island, 48 in Manhattan, and 74 in Brooklyn—for a total of 141 vessels. On that same typical day, we would find only 17 vessels docked in Newark.[32]

Despite these discouraging numbers, Port Newark was kept in the game by constant dredging of both the Passaic River and Newark Bay. In 1926 federal appropriations were used for dredging the Passaic River above the city of Newark. The 1927 appropriation was spent on the lower river and Newark Bay.[33] As a result of this activity, the Newark Bay Channel was maintained at a depth of 30 feet at mean low water, with a width of 400 feet. The Port of Newark Terminal Channel, where the shipping facilities were located, also had a 30-foot depth at mean low water, with a width of 685 feet. Meanwhile, a depth of 20 feet was maintained for the lower Passaic River.[34]

The Great Depression Years

The stock market crash of October 1929 shocked the entire nation. The resulting Great Depression was severe, and its effects were felt worldwide. The city of Newark, heavily dependent on industry, was especially hard hit. The misery in Newark was well documented by historian John T. Cunningham: "More than 7,500 families—about 37,500 persons—were on direct relief in the middle of 1930. The poor stood in long lines hoping to see the Overseer of the Port in his basement office in City Hall.... Men sold apples in the streets, shoveled snow, ran errands, or openly begged."[35]

As personal income fell and prices dropped, international trade fell by more than half, with devastating effects on all the ports of the region, including Port Newark. It was sad to see once-prosperous businesses fall by the wayside. One such case was that of Transmarine Lines, which had been launched in 1921 with Newark as its home port. Using a fleet of cargo ships, built by the Submarine Boat Corporation discussed earlier, Transmarine had provided regularly scheduled service to Houston, Texas, as well as to Los Angeles, San Francisco, and Oakland. But the company went into the hands of receivers in January 1930. By June 1931, the company's twenty-two remaining ships were sold to the Portland-California Steamship Company for $400,000, or slightly more than one-fortieth of their original cost.[36]

To offset the widespread unemployment, the federal government launched a number of programs to aid the infrastructure and to provide jobs. For example, in 1931 the hard-pressed federal government appropriated emergency funds for rivers and harbors, but it was not nearly enough. Of the $22.5 million allotted in the crisis, only $434,000 came to the New York–New Jersey region, and only a fraction of that was available for work on Newark Bay as well as the Hackensack and Passaic rivers. The Port Authority estimated that, at the slow rate of appropriations, it would take thirteen years to properly dredge the region's waterways.[37]

Things gradually got better. Most economic historians point to 1933 as a turning point for the nation, though the recovery at Port Newark was painfully slow. By the late 1930s, there was a trickle of good news here and there, though nothing like a full recovery. For example, in late 1936, the City of Newark leased eighteen acres at Port Newark to Swift and Company for fifty years in exchange for the company agreeing to build a $3 million vegetable shortening and edible oil plant. In time, the plant was expected to provide 1,000 permanent jobs. A spokesman for the company praised the facilities, adding that Port Newark was a "logical location, with natural advantages as a land, water, and air distributing center."[38]

The Lighterage Problem Revisited

In March 1929, New Jersey officials reopened the old lighterage case, discussed earlier. As before, New Jersey wanted the ICC to force the railroads to charge less for goods delivered to the New Jersey side of the harbor. New Jersey petitioned for the abolition of free lighterage. Raymond J. Dempsey was the superintendent of Port Newark, and he argued his case forcefully before examiner Earl W. Steer at the hearing in Newark. Dempsey claimed that free railroad freight lighterage across New York Harbor was choking the life out of Port Newark.

Dempsey went on to say that the city's annual income from shipping was about $200,000, while it had spent some $15 million on the port. There was no way, he said, that Newark could get a fair return on its investment unless the ICC granted a lower freight rate structure for New Jersey. Dempsey pointed out that Port Newark was using less than 25 percent of its facilities and that he advised the city to avoid any further construction until the present facilities were fully utilized.

Dempsey was vigorously opposed by John F. Victory, assistant corporation counsel of New York City. Victory pointed out that Port Newark had enjoyed substantial increases in total tonnage handled. Dempsey agreed that the tonnage had indeed gone up, but such statistics were misleading because the bulk of the tonnage was in lumber, a cheap commodity. The City of Newark was not satisfied with Port Newark serving as a mere lumber terminal.[39]

In 1933 the ICC examiner supported the New Jersey position, rejecting New York's argument that the freight rates should be unified throughout the port. It was a resounding victory for New Jersey and a stunning defeat for New York. The examiner wrote that free lighterage had hurt New Jersey's economic development and that rail lines floating goods across the Hudson would have to start charging extra for the service. Things looked good for Port Newark, but in July 1934 the full commission reversed the examiner, supporting the free lighterage system once again. New Jersey officials were furious, having their short-lived victory snatched away.

New Jersey did not give up the fight, and a new round of hearings took place in 1939–1940. The hearing was conducted by ICC examiner Burton Fuller at the Hotel St. George in Brooklyn. The arguments were heated because the stakes very high. On the New York side, there was great fear that charging for lighterage would make their side of the port less competitive. Iron and steel might be diverted to Baltimore, sugar and green coffee to New Orleans, corn and domestic grain to Philadelphia.[40] Meanwhile, the attorney for New Jersey, Milton P. Bauman, said that free lighterage "was a racket since its inception." He then went on to say, "Lighterage conditions are no longer a racket: they have been sanctified. Now they are just unlawful."[41] Despite Bauman's eloquence, once again, the ICC supported free lighterage. Exhausted and frustrated, New Jersey officials finally gave up the fight. Despite all the money that had been invested in Port Newark, city fathers began quietly talking about abandoning the port altogether.[42]

World War II Years

After the declaration of war by the United States against Japan on December 8, 1941, the U.S. Navy embarked on a crash program of shipbuilding. At first Port Newark was overlooked because the Submarine Boat Company shipyard left over from World War I was in bad shape. But given the desperate need for more ships, the federal government saw potential in Port Newark; and in January 1942, it purchased the old shipyard and agreed to spend some $12 million to convert it into a new, modern steel facility. The plant was to be owned by the Navy but was to be operated by the Federal Shipbuilding and Dry Dock Corporation of Kearny, New Jersey, a subsidiary of the United States Steel Corporation. The site, consisting of 112 acres, with 28 ways, or inclined structures used to support ships while being built before launching, had a good deal of potential.[43]

Suddenly, the sleepy seaport became a bustling shipyard turning out dozens of warships. Thousands of workers were hired. They were quickly at work simultaneously fixing up the place and beginning work on new vessels. Fabricating and shaping the steel began in July 1942, and the first keel was laid on August 10. Four more keels were laid on September 5, and two auxiliary ships were launched on October 10. Because of the acute labor shortage, the shipyard began hiring women as mechanics and riveters. By October 1943, there were 19,503 employees, working around the clock in three shifts. All in all, they built seven destroyers, fifty-two destroyer escorts, and seventy-eight landing ships.[44]

Of all the ships built at Port Newark during World War II, the seven destroyers were the most impressive. With their high speed and substantial weapons, they were designed to defend other ships against enemy attacks. They could help to defend either naval fleets or merchant convoys. They were fairly big, at some 390 feet in length, and very fast, with a maximum speed of some 36 knots. They had considerable firepower, equipped with six naval guns, capable of firing projectiles five inches in diameter, useful against both enemy surface ships and aircraft. As multipurpose ships, they were also equipped with radar, sonar, antiaircraft guns, depth charges, and torpedoes.[45]

Although the Port Newark shipyard took great pride in building its seven substantial destroyers, it was better known for building large numbers of smaller, lightly armed ships called destroyer escorts, or DEs. They were built specifically for the Battle of the Atlantic, where German U-boats were sinking allied ships at an alarming rate. What was needed to defend against the submarines was a large number of smaller and cheaper destroyer-like vessels modeled on the British and Canadian corvettes. To guard a slow-moving convoy did not require a fast-moving destroyer. The DEs could travel at a rate of only about 24 knots, but it was good enough. They had a tremendous advantage in maneuverability. A typical DE had a narrow turning circle of only 400 yards, compared to the 880 yards required for a destroyer. This was a tactical advantage in chasing the wiley submarines. Most importantly, a DE could be built in less than half the time of a fleet destroyer and at a third of the cost.[46]

The DEs built at Port Newark served in both the Battle of the Atlantic and later in the Philippine campaign and the island-hopping advance against Japan. The most decorated of these ships was the USS *Bronstein* (DE-189), launched at Port Newark on November 14, 1943. The ship was deployed to the Atlantic Ocean, where it was tasked with providing escort service against both submarine and air attacks to protect both naval warships and merchant marine convoys. Unusually successful in antisubmarine action in the period of February and March 1944, the ship received the Presidential Unit Citation for extraordinary heroism against an armed enemy of the United States.[47]

At the end of the war the Navy was downsizing. There were more than enough warships for peacetime patrols, so the shipyard was closed, never to reopen. Eventually, all of those hastily built DEs were either given to foreign navies or sold for scrap. But one of them lives on, at least in the popular imagination. That would be the USS *Eldridge* (DE 143), the subject of the legendary Philadelphia Experiment. According to the story, the *Eldridge* was the ship used in a military experiment carried out at the Philadelphia Naval Shipyard on or about October 28, 1943. The idea was to render the ship invisible to the enemy, which would have been an astounding military advantage. Instead, the experiment went terribly wrong. The ship and its crew were supposedly teleported into another dimension, traveling through time and space. It was said that most of the crew never returned, and the few who did came back insane.[48]

It must be said that the Philadelphia Experiment is an intriguing story. In a wartime atmosphere of top-secret military experiments, the story has just enough plausibility to attract a devoted core of believers. It seems to be a shame to spoil a good story; however, there appears to be no solid evidence to back it up. The Department of the Navy has released an official document on the Philadelphia Experiment, refuting the story, point by point. The report begins by stating that "no documents have been located which confirm the event." Even more conclusively, the report says, "The Office of Naval Research (ONR) has stated that the use of force fields to make a ship and her crew invisible does not conform to known physical laws."[49]

Of course, students of conspiracy theories know that official denials mean nothing to true believers. For them, it is just further evidence of a coverup. The story as an urban legend has had amazing staying power. In 1984 it was made into a Hollywood movie, *The Philadelphia Experiment*, starring Michael Paré and Nancy Allen. In this version, two of the sailors are teleported into the future. They find themselves in the Nevada desert in 1984. A nurse called Allison Hayes, played by Nancy Allen, assists the sailors. In time, there is a love affair between Allison Hayes and sailor David Hertig, played by Paré.[50] The story has a firm grip on the popular imagination and has been told repeatedly. As recently as 2012, there was a made-for-television version of the original film that appeared on the Syfy channel. Michael Paré also appeared in this version, but in a different role, as a secondary antagonist.[51] This strange story of teleportation may be a hoax, but that has not stopped it from becoming a staple of our popular culture.

2

The Post–World War II Era

● ● ● ● ● ● ● ● ● ● ● ● ●

In the post–World War II period, the city commissioners of Newark had a full plate of problems. There were racial tensions, which led to white flight to the suburbs. Employment was slipping, and property values were declining, eroding the tax base. Much of the existing housing stock failed to meet minimum standards of health. The city hospital could not keep up with its patient load. The streets, which had been neglected during the war, were in need of $10 million worth of repairs.[1]

Meanwhile, the city was also responsible for the operation of the seaport, which had opened in 1917, and the airport, dating to 1928. Both were sources of civic pride, but both had been financially disappointing. To really succeed, both the airport and the seaport would need a huge infusion of capital—an unlikely scenario give the city's other pressing needs. Where would the money come from? At first, it looked like an impossible problem. However, by happy coincidence, this was a time when the Port Authority was looking for new projects. Negotiations would be delicate. Clearly, the City of Newark was in desperate need of resources, but the politics were tricky. The Port Authority did not want to appear to be unduly powerful and acquisitive. The City of Newark did not want to appear to be giving away its sovereign assets. Thus, there was a good deal of talk and compromise behind closed doors and out of the public spotlight.

Historian Jameson Doig has explained that the Port Authority developed a carefully crafted takeover strategy whenever going after municipal assets. Never wanting to appear greedy, the agency would develop a takeover plan only if it first received a request from the affected municipality. Once the request was received, the agency would gather up a panel of experts who would be charged with studying the problem and producing a report. Then the report would be forwarded to the local government for its deliberation. It would be up to the municipality to debate the issue and make a decision. Only if asked would the Port Authority proceed to implement the plan. This approach required considerable behind-the-scenes maneuvering and a good deal of patience. Let us see how this strategy played out in the acquisition of Port Newark.

A key figure in this unfolding drama was Harland Bartholomew, a civil engineer trained at Rutgers University. Over the years, he had built up a reputation as the dean of comprehensive city planning in the United States, and he formed the consulting firm of Harland Bartholomew and Associates. This firm was called in by the City of Newark in 1943 to study the ongoing needs of the city in terms of physical growth and financing. The firm prepared exhaustive reports on population trends and housing conditions. When Bartholomew's planners turned their attention to the airport and the seaport, they found both of the facilities undercapitalized and saddled with many patronage jobs. Some of these were, in fact, "no-show" jobs. It seemed to the planners that it would be unwise for the city to continue to run the airport and the seaport as regular city departments. Now it was a matter of persuading the city commissioners to accept the findings of the planners.[2]

Bartholomew himself visited Newark in the summer of 1944, and he pondered the problem of adequate funding for the airport and the seaport. He took the obvious step of reaching out to Austin Tobin, executive director of the Port Authority. To Bartholomew's pleasant surprise, he got an appointment with Tobin right away. Meeting in New York, Bartholomew laid out a plan for Port Authority takeover of the two facilities; he found Tobin quite receptive. Although Tobin was more interested in the airport than the seaport, he expressed a keen interest in leasing the two facilities for a forty-year period. Returning to Newark, Bartholomew was giddy with excitement, but he knew that these conversations would have to be kept confidential. In order to make things work, it would have to appear that initial overture would be a request coming from the City of Newark asking the Port Authority to prepare a proposal.

Bartholomew asked William Anderson, one of his associates, to prepare a detailed report on just how much it would cost to rehabilitate and modernize the airport and the seaport. The report was completed by October 1945. The report pointed out that the City of Newark, over the years, had spent $9.6 million for Port Newark and $7 million for the airport, with insufficient returns on such a heavy investment. Besides, the city was faced with enormous other expenses for schools, hospitals, and streets. The report concluded that the Port Authority had the expertise and the resources to best handle both facilities. The report was submitted to the Newark Planning Board, made up of both civic officials and business executives. By early December 1945, the Planning Board endorsed the report and urged the Port Authority to meet with Newark's elected officials to purse the takeover.

The process was neither easy nor quick. The city commissioners had good reasons not to like the plan: it would hurt the city's pride and would cost many patronage jobs for friends and party workers. The commissioners were understandably divided in their opinions. On the one hand, vice chairman Joseph Byrne, taking the long view, argued that the facilities should be placed under Port Authority control. On the other hand, commissioner John A. Brady was vigorously opposed, arguing that it would be great loss for the city. In their hearts, most of the commissioners would have liked to continue business as usual, but they were under considerable pressure from the business leaders in the community to accept this bitter medicine. Finally, on December 27, 1945, the city commissioners asked the Port Authority to study a takeover. Immediately, the Port Authority agreed, but the matter was far from settled. A study was a good first step, but ahead lay a long period of delicate negotiations.[3]

By late July 1946, the study was completed, and Austin Tobin was prepared to make a public presentation to Newark's mayor and commissioners. He said that the Port Authority could assume responsibility for both the airport and the seaport. He promised to modernize and improve the facilities. The package included $11 million to be invested in Port Newark, and $55 million in Newark Airport. In addition, the city would receive $1 million every year for five years, and afterward $100,000 a year in place of taxes. In other words, the city stood to gain some $7.5 million over a thirty-year period. The plan was to invest enough capital up front in order to make both the airport and the seaport eventually self-sufficient.

The plan received a mixed reaction. While the region's newspapers and business leaders were enthusiastic, all four city commissioners were adamantly opposed. There was still hope because the mayor, Vincent J. Murphy, was receptive to the plan. Meanwhile, Austin Tobin kept a low profile, figuring that eventually the city commissioners would realize that they simply did not have the revenue to fix up their obsolescent facilities. The Port Authority's strategy was to be patient, rather than to push for a quick victory.[4]

In early September 1946, the Port Authority increased the pressure on the city commissioners by sweetening the offer. The agency announced an "alternative method" that amounted to the payment of $9,202,500 over a thirty-year period, a considerable increase over the original offer. In addition, the agency agreed to take over the cost of street maintenance in both the seaport and the airport, for a savings to the city of some $30,000 to $40,000 a year. Last, the agency agreed that the properties would continue to be called Port Newark and Newark Airport.[5]

Meanwhile, Austin Tobin was working behind the scenes to line up support for his takeover plan. In November 1946, Alfred E. Driscoll was elected as governor of New Jersey. Without wasting any time, Tobin gathered up his New Jersey board members and paid a visit to Driscoll in Trenton, explaining the advantages of Port Authority management of the two Newark facilities. By January 1947, Driscoll was solidly behind the plan. He made his opinion quite clear in his first address to the legislature: "The time has come when the people of that city can no longer be expected to underwrite huge capital investment outlays of this kind."[6]

Pressure kept building up on the city commissioners over the first few months of 1947. The plan now had the backing of the governor as well as most business leaders and editorial writers. In early October 1947, the Broad Street and Merchants Association reaffirmed its approval of the deal. It urged the City Commission to go ahead and sign the contract: "Too much valuable time has already been lost." The statement said that the proposed lease was "advantageous to the city." It dismissed the objections of the holdout commissioners as invalid. Finally it said, "Within the limits of our ability to analyze the complicated data and estimates involved, we are satisfied that while not ideal from every standpoint, the lease would improve substantially the city's financial picture, and that every individual taxpayer in Newark would benefit from the conversion from what is now a financial burden to a source of substantial income."[7]

Notwithstanding the strong support for the takeover plan by local merchants, there were still powerful vested interests that were opposed, most notably the railroads. William J. Clancy was the attorney who represented the Pennsylvania and Lehigh Valley railroads. He said, "Experience has shown that when the Port Authority takes over an activity that competes with private business, their competition is so unfair that it drives competition out of the city." He went on to say, "The Port Authority with its sovereign prerogatives of eminent domain and exemption from taxes can quote rates that are far below those that can be quoted by private capital. Thus, while the Port Authority flourishes, the city's lifeblood, private industry, is forced to leave. This reduces the freight revenue of the railroads."[8]

As the final vote of the city commission approached, Governor Driscoll intervened. He pointed out emphatically to the commissioners that he had been sympathetic to providing more state aid to the older cities, including Newark, but that now was the time that they had to help themselves by cutting off facilities that were losing money. Finally, on October 22, 1947, with the combined pressure of the merchants, the press, and the state, the City of Newark and the Port Authority signed an agreement that would lease the airport and the seaport to the bistate agency for fifty years. The way was now clear for the Port Authority to open its checkbook and to put Port Newark on the map as a major destination for seaborne commerce.[9]

In 1948, the Port Authority could forge ahead with its entrepreneurial vision for Port Newark. The first step was to acquire more land and buildings. In July the agency acquired two separate properties. The first was a purchase from the War Assets Administration of 38 acres of land and buildings of the former Sears-Roebuck area, on the south side of Port Newark, for $1,016,000. This property included six warehouses with 250,000 square feet and a five-story building of 110,000 square feet. The second purchase was for buildings and equipment on 7.75 acres of the Franklin Lumber Company, fronting on the Port Newark channel, for $200,000.[10]

Continuing the expansion of Port Newark, the agency signed a lease agreement on September 28, 1948, with the U.S. Navy for 33 acres of land and buildings on the northeast channel. The leased area included seven deep-sea ship berths, two dockside transit sheds, and two warehouses. The lease was to run for ten years and be renewable for five more. In place of rent, the authority would invest some $377,000 in rehabilitation and new construction, to be returned to the Navy after fifteen years. The two transit sheds had 63,000 square feet of floor space. The smaller of the two warehouses had 40,000 square feet; the larger warehouse, 160,000 square feet. Both warehouses had tracks connecting with the Jersey Central, Pennsylvania, and Lehigh Valley railroads.[11]

Keeping its promise to develop Port Newark vigorously, the agency announced in November 1948 that it planned to invite bids on an issue of $7 million of marine terminal bonds. The plan was to issue thirty-year term bonds, secured by terminal revenues and by the general reserve fund.[12] In

January 1949, the Port Authority issued its annual report for 1948. That report celebrated the success of increasing the trade through the Port of Newark. The increase was accomplished by encouraging new types of cargo. For example, new imports included 2,000 tons of African vermiculite used for insulation, some 4,400 tons of chrome ore from Africa, and a quantity of steel received from Sweden. New exports included 1,250 tons of scrap metal sent to the Philippines and 718 drums of oil shipped to Finland. The report boasted that buildings and rail tracks had been rebuilt. In addition, the fender system around wharfs and bulkheads had been repaired. It was also announced that plans were under way for dredging the channel to a 35-foot depth to accommodate bigger ships.[13]

As time passed, it became clear that the city commissioners, however reluctantly, had made the right choice in turning management of the facility to the Port Authority. The agency, in turn, wasted no opportunities to boast about its accomplishments at Port Newark. For example, the Port Authority on May 24, 1950, cleverly timed the dedication of two newly built cargo terminal buildings to coincide with World Trade Week, an annual event always held the third week of May. The Port Authority hosted a gala event for some 1,200 guests, including leaders from the shipping industry. The guests were given a boat tour of the harbor and a luncheon, followed by the dedication ceremony for the new cargo terminals. Of course the press was also invited. Port Authority officials told the assemblage that tonnage handled at Port Newark in 1949 had shown a significant increase over the previous year.[14]

Keynote speaker at the ceremony was Howard S. Cullman, chair of the Port Authority, who cited the new buildings as examples of what the agency could do. He said that the prosperity of the region depended on a port that was "properly promoted and operated on a business-like basis." The new cargo terminals were state-of-the-art, built at a cost of $2.5 million. Each was designed with a 40-foot apron, the paved area between the waterfront edge and the cargo terminals, on the shipside to receive cargo. The cargo could then be sorted and temporarily stored inside the cavernous building. Then the cargo could be shipped out from the rear of each building where there was a 12-foot platform for trucks and trains.[15]

The Korean War

On June 25, 1950, the North Korean Army invaded South Korea. The United Nations promptly condemned the invasion, demanded that the invading forces withdraw, and urged UN member states to aid South Korea. On June 27, U.S. president Harry Truman authorized the use of American military forces in Korea. A coalition force from fifteen member nations was placed under the command of the charismatic general Douglas MacArthur.[16] The Korean War has often been called "the forgotten war," but we cannot overlook its important role in our story.

In the first phase of the war, the ill-equipped American troops from Japan were driven back. Only later did the slogging war on the ground turn around when adequate amounts of equipment reached the hard-pressed troops. To meet the urgent need to supply the troops with ammunition, vehicles, food, and medical supplies, the U.S. military had to rely on the logistics lessons learned in World War II. Large numbers of cargo ships were needed in a hurry. This meant quickly reactivating many of the Liberty and Victory ships, which had been mothballed after the war. The Liberty ships had been the workhorses of World War II. They were produced in large numbers to replace the losses caused by German submarines. Based on a prewar British design, with reciprocating steam engines, they were rather slow, maxing out at 11 knots, but they were available in large quantities.[17] Built quickly and easily, the Liberty ships were easy targets for armchair critics who lambasted them for their slow speed. Rear Admiral Emory Scott Land, a staunch advocate of the Liberty ships, known for his salty remarks, replied, "If you want fast ships, fast shipbuilding, fast women or fast horses, you pay through the nose for them."[18]

In 1943, even as shipyards were turning out large numbers of Liberty ships, the Maritime Commission was planning for new and faster type of cargo carrier. Plans called for the construction of the improved Victory ships, evolved from the Liberty ships, but slightly larger, with modern turbine engines. The hull was less box-like and more streamlined. With better engines and better hull design, the ships were capable of making some 15 knots. The extra speed made these ships less vulnerable to U-boat attacks.[19]

The preferred cargo ship used during the Korean War was the so-called C-2 type, which differed from the famous Liberty and Victory ships used during World War II. The C series vessels were more modern and were designed to be commercially viable, with remarkable speed of some 15 knots and good fuel economy. Each of these cargo ships had five holds, or large lower compartments for the stowage of cargo. The holds were numbered one to five, running from bow to stern. Each hold could be accessed by means of a hatch, or opening on the ship's deck. Dimensions of the hatches were an ample 20 by 30 feet, allowing for the passage of most cargo with the exception of oversize items such as locomotives or naval guns, which had to be stored on deck.

The C-2 type ships were largely self-sufficient for loading, with cargo handling gear consisting of fourteen 5-ton cargo booms and two 30-ton booms at hatches 3 and 4.[20]

Today, very few people remember that Port Newark played a key role in the military supply chain for the Korean War. Ships of all types were pressed into service, including the old Liberty ships, along with the newer Victory ships and C-2 cargo ships. The story of how ships were loaded with cargo in those days is particularly interesting because the method used, known as break-bulk cargo, can be traced back to the very beginning of the history of shipping. Workers loaded the ships with bags, boxes, crates, drums, and barrels. Sometimes the goods would be bundled into unit loads secured to a pallet or skid. No one knew at the time that the labor-intensive days of handling break-bulk cargo were numbered. In a few short years, the adoption of the modern shipping container would change everything. Moving cargo on and off ships in containers would be more efficient and would reduce theft and damage.

We can best understand the historical significance of containerization by taking a close look at the way things were done earlier. In this regard, I had the good fortune to get a firsthand account from Frank Greco, who explained the process patiently. Most of the cargo would arrive at Port Newark in advance of the arrival of the ship; the cargo would then be stored in a nearby warehouse, protected from the elements. When the ship finally arrived, dockworkers would take the cargo from the warehouse to the dock, where it would be lowered into the cargo holds of the ship by means of the ship's derricks or cranes. Then it would be the job of the marine carpenters to secure, support, and protect the cargo from shifting during the voyage by means of lumber bracing, called dunnage.[21]

Frank Greco worked as a marine carpenter in Port Newark from 1951 to 1953, and he shared with me a detailed account of how the work was done. Considerable care had to be taken to avoid any damage to the goods. Typically, the first goods to come aboard the ship would be boxes and crates and barrels, mostly with foodstuffs, canned goods, and beer. This cargo would go on the lower deck, and it would have to be secured to prevent any movement. In addition, the cargo would have to be kept away from the ship's sides, to prevent contact with the moisture formed on the steel surfaces. Packing crates and cases of medium size could usually be loaded only in stacks of five similar items, to avoid crushing. Once the first layer of cargo was stowed and secured, a platform of wooden dunnage would be placed on top of it.

Then the marine carpenters could proceed with the second layer, which might consist of vehicles such as Jeeps and small ammunition trailers. The trick here was to use dunnage to distribute the weight and not crush the goods below. With sufficient care, it could be done. A typical U.S. Army Jeep weighed just over 2,000 pounds, and a typical M-10 U.S. Army ammunition trailer weighed about the same. Both the Jeeps and the trailers would have to be secured with cribbing, or wooden bracing, all the way around. The marine carpenters would use 4-by-4 inch or 4-by-6 inch timbers right up against the tires on each side, and in front and back. Once that was done, an additional crew would come in with cables or wires and secure the vehicles to cleats that were welded onto the bulkhead, or wall of the ship. The cables could then be adjusted and tightened by means of turnbuckles. Everything was tied down and interlocked to prevent shifting of the cargo at sea.

When the cargo in a given hold was fully loaded, the next step would be to close the hold with a heavy wooden hatch cover. The hatch cover would then be covered with a large waterproof tarp, with extra material on all sides. The extra material would then be fixed to the deck by means of steel bars that were bolted down. Once covered, the hatch would be pitch dark. However, an Army inspector would go down there with a flashlight, just to make sure that everything was tied off and interlocked. The Army was not taking any chances with shifting cargo.

There was an interesting division of labor on the docks at Port Newark. The dockworkers were responsible for loading the goods onto the ship, and the marine carpenters were responsible for securing the goods onboard the ship. The dock workers were paid by the hour. Left to their own devices, they would not hurry. However, the contractor who hired the dockworkers was paid by the piece, so he would constantly push the dockworkers to work quickly. As a result, the marine carpenters, who were also paid by the hour, had to hustle, perhaps against their will, to keep up with the incoming cargo. If they did not hustle, they could be replaced, so they learned to work very quickly.

Most of the work consisted of cutting and fitting pieces of timber for bracing. For the smaller jobs, the marine carpenters might work with 2-by-4 or 4-by-6 lumber. However, for big jobs, they might need 6-by-8 or even 12-by-12 wood. In any event, the tool of choice was the portable circular electrical saw. Handsaws would have been much too slow. Normally such saws run on 110 volts of power, but this gang was not satisfied with normal speed. Therefore, they went to the fuse box on the ship and went with two circuits at once.

Now the saws were running on 220 volts. Of course, the saws did not last very long, but they were very powerful while they lasted. Frank said that you could go through a 12-by-12 timber as if it were a piece of paper. To further increase the speed of the work, they tied off the guard. It was another unsafe practice, but it allowed them to increase the pace of the work.

The workplace was a madhouse. One worker might have a ruler and be taking measurements and yelling them out to the man working the saw. Once the timber was cut to order, another couple of workers would nail it in place. Meanwhile, the dockworkers would keep dropping in Jeeps and trailers with no letup. The work was particularly hazardous for the man operating the saw because there was always the concern that once you sawed off a given timber, it might fall on your foot. Frank had his big toes smashed twice. I asked him why he never wore steel-tipped boots. He explained that they were not allowed on the job because the timbers were very heavy. Once a timber would fall on a steel-tipped boot, it would cut off the toes of the wearer. With regular shoes, your toes would only be smashed, not cut off.

Despite all of the job's hardships, Frank loved it. It paid well and there was plenty of overtime. The hours were often irregular. For example, if a ship came into Port Newark at ten o'clock one morning, that was when the workday began. In such a case, one might reasonably expect to go home at five o'clock and continue the job the next day, but sometimes the workers were required to stay and work through the night. It was tough, but the pay was good. In the two years that he worked at Port Newark, Frank could recall only two times that he worked a straight forty-hour week. That was when there were no ships to load. Instead, the marine carpenters were put to work on "dunnage reclamation." In other words, they had to remove the nails from old, used dunnage, sort the timbers by size, and stack them neatly for reuse.

As the summer of 1953 approached, an old-timer took Frank aside and said, "Just remember that when there's a war going on, there will always be plenty of work. But when the war's over, the work will be over." Sure enough, the Korean War did end a few weeks later with the signing of an armistice agreement on July 27, 1953. Therefore, Frank Greco took the old-timer's advice and turned in his resignation.[22] Meanwhile, the wartime boom at Port Newark slowly wound down. However, the underlying fundamentals were sound, and the Port Authority continued to invest in improvements.

For example, just a few months following the end of the Korean War, the Port Authority opened a new large marine depot at Port Newark. Built at a cost of some $6 million, the new facility was occupied by the Waterman Steamship Corporation, an American deep-sea ocean carrier. The large project had 1,500 feet of new wharfage and a series of three connected transit sheds with 270,000 feet of covered cargo space. The facility was capable of handling about 600,000 tons of cargo annually. In addition to the cargo space in the transit sheds, the terminal had office space, a passenger waiting room, locker rooms, customs offices, and special rooms for sensitive goods.[23]

The dedication of the new facility took place on Thursday, March 25, 1954. More than a thousand people attended, including shippers, importers, and exporters and officials from the Port Authority, the city, and the state. The keynote speaker was Governor Robert B. Meyner, who said, among other things, "We must buy the products of a revitalized Europe if Europe is to be able to buy from us. Here world trade will help all of humanity. It is a mistake to believe that development of Europe and the Far East will shrink our foreign markets. On the contrary, it will expand them."

The opening of the new facility certainly represented a victory for Newark mayor Leo P. Carlin since the Waterman Steamship Corporation was moving from Brooklyn to Newark. However, Carlin did not dwell on Brooklyn's misfortune with smugness or malignant pleasure. The mayor simply reviewed the history of the port, which had been leased by the Port Authority in 1948. He said that the occasion "commences a new era in sea transportation in New Jersey—envisioned by far seeing men back in those early days of 1917 and 1918, but slow in reaching fruition."[24]

The 1950s seemed to represent a sudden growth spurt for Port Newark. The opening of the Waterman terminal in March 1954 was followed by the announcement of another expansion in November 1955. The Port Authority declared its intention to construct a four-berth pier facility on a 24-acre tract adjacent to the Waterman terminal. This new terminal was to be leased to Norton, Lilly and Company, ship agents since 1841. (Not to be confused with Eli Lilly and Company, the global pharmaceutical company.) This company represented a number of ship operators, who found that it was more efficient to use an agent rather than establish an office at every port. The agent would handle paperwork, customs, scheduling for arrivals and departures, and so on.

The new facility was expected to handle an annual volume of 400,000 tons of around-the-world general cargo. It was to have a wharf area 2,400 feet long, two transit sheds, 880 feet long and 200 feet wide as well as a two-story air-conditioned office building of about 8,000 square feet, and a 400-car employee parking area. The sheds on the landside were to have a truck-loading area, served by a platform capable of handling 140 trucks at one time. On the water side there was to be an apron with double railroad tracks.[25]

The Norton, Lilly and Company terminal was announced on Wednesday, November 30, 1955. Just two days later, there was another blockbuster announcement: New Jersey governor Robert B. Meyner revealed that the Port Authority would acquire a 450-acre tract of privately owned tidal marsh in the city of Elizabeth adjacent to Newark Bay, just south of Port Newark. The Port Authority set out to acquire the land, much of it owned by the Central Railroad of New Jersey. To finance the purchase and start construction, the agency set aside $4 million. The plan for Port Elizabeth called for accommodating simultaneously twenty-five full-sized vessels and handling 2 million tons of cargo annually. Governor Meyner said, "In the interest of sound financing, both the channel and the port structures will be built gradually as the demand requires. We are going to turn this unused marshland into one of the most important port areas in the world."[26]

The announcement of the development of Port Elizabeth was a bold move, a game-changer. Meyner's timing could not have been better. A number of factors came together to make this the right move at the right time. First, Port Newark had simply run out of room to expand. Second, there were no environmental objections at this early point in time to developing "unused marshland." Third, there was growing public frustration with the waterfront corruption on the New York side. The 1954 movie *On the Waterfront* was on everyone's mind. The film had dramatized the violence, corruption, extortion, and racketeering on the waterfronts of Manhattan and Brooklyn.[27] There was hope that the Waterfront Commission of New York Harbor, established in 1953, could do a better job at Port Elizabeth, with a fresh new start. Fourth, and most important, this brand-new port was being developed just in time to take full advantage of the new system of containerization—the subject of our next chapter.

3

The Invention of Containerization

● ● ● ● ● ● ● ● ● ● ● ● ● ●

The development of containerization was a major change in transportation technology—an innovative idea with profound consequences for world trade. It all began at Port Newark on April 26, 1956. It was a chilly day, with a high temperature of 48 degrees and intermittent showers. On a dock outside shed 154, some fifty-eight 33-foot-long containers were swung aboard the SS *Ideal X*, a converted World War II oil tanker stretching 524 feet, with the previous name of *Potero Hills*. Previously, oil tankers would carry a profitable load of oil from Houston to Newark, but they would make the return trip with only ballast, something heavy to improve the stability of the ship. Now tons of cargo could be carried on the southbound trip, making both legs of the voyage pay. A huge crane placed the containers (they were understandably called "trailers" at that time) on a metal platform of the tanker, above the pumping machinery.

Attending the event were Donald V. Lowe, chairman of the Port Authority of New York; Austin J. Tobin, the Port Authority's executive director; and Newark mayor Leo P. Carlin. In his speech, Carlin addressed the inventor of the concept, Malcom McLean: "With the first sailing of a container ship, you will have converted a challenging, revolutionary transportation concept into a practical reality."[1]

FIGURE 1 The *Ideal X* under way with a focus on the starboard bow. The ship was a converted World War II oil tanker, said to be the first commercially successful container ship. The ship was the brainchild of Malcom McLean, who realized that converting a surplus tanker was cheaper than building a brand-new, purpose-built container ship. (Courtesy of the *Journal of Commerce*.)

It was not necessary to put the whole trailer on the ship. Instead, everything underneath each trailer would be removed to save space. The trailers would be removed from their steel beds, axles, and wheels. All that was left was the trailer bodies, and these could be stacked. At the end of the trip, the trailer body could be lowered onto an empty chassis and driven away to its destination.[2] Later that evening, the ship left for Houston, where it arrived six days later at City Dock 10.

The average loading time was seven minutes per container. The whole ship was loaded in less than eight hours. Under normal circumstances, the freight on the ship would have gone to Houston in over-the-road trucks—a trip of about six days. It would take the ship the same amount of time, but there would be a tremendous savings of about 30 percent.[3] It was a breakthrough—the brainchild of one man, Malcom McLean. How did he come up with such an original idea? In his favor was the fact that he had no experience in shipping. He was an outsider, a truck driver.

To understand McLean's thinking, we have to go back in time to his early life in rural North Carolina. He grew up as a farm boy, one of seven children. The son of a farmer who also worked as a mail carrier, he early on learned the value of hard work. His mother began raising a few chickens, sharing a dozen

FIGURE 2 An aerial view of the *Ideal X* carrying fifty-eight containers from Port Newark, New Jersey, to the Port of Houston, Texas. At the time, this voyage received very little media attention. Now, years later, historians recognize it as an event of huge importance, launching a revolution in global trade. (Courtesy of the *Journal of Commerce*.)

or two with family and friends. Gradually, she began reaching cash-paying customers. Young Malcom helped by acting as salesperson, earning a small commission. Graduating from high school in 1931, during the depths of the depression, he had no money for college, so he took up work at a local service station pumping gas. Life was a struggle, but before long he had saved up enough money to buy a used truck for $120. His career in transportation was launched, and he began calling himself the McLean Trucking Company.[4]

McLean got started by hauling topsoil, manure, produce, and livestock for his hometown farming community of Maxton. He soon gained a reputation for reliability, and he acquired five additional trucks with a team of hired drivers. This move enabled him to stop driving himself and to start reaching out to new customers. For a while, he seemed to prosper; but, as the depression deepened, he lost customers and had to resume driving. Things were bleak, and it looked like he might lose his business altogether.

To survive, McLean started accepting long-haul contracts, requiring him to be away from home for days at a time. In November 1937, just before

Thanksgiving, he got a job hauling cotton bales from Fayetteville, in southeastern North Carolina, to Hoboken, in northern New Jersey, then a major transatlantic port. After a long drive, he arrived in Hoboken, joining a long line of trucks waiting to be unloaded at the pier at the foot of Third Street. The mixed cargo was being placed, piece by piece, aboard the *Examelia*, an America Export Lines ship that would take everything to Istanbul. The line moved slowly, and there was not much to do. He had to wait hours for his truck trailer to be unloaded. Years later, he recalled, "I had to wait most of the day to deliver the bales, sitting there in my truck, watching stevedores load other cargo. It struck me that I was looking at a lot of wasted time and money. I watched them take each crate off the truck and slip it into a sling, which would then lift the crate into the hold of the ship."[5]

This oft-quoted sudden realization was the birth of an idea that would take a full nineteen years to blossom into fulfillment. During that time, McLean focused on his trucking business. However, his perceptive, intuitive thinking separated him from others in that business. He had plenty of time to think and plan. By the early 1950s, McLean had built his company into the largest trucking fleet in the South, with 1,776 trucks and 37 transport terminals up and down the East Coast.

Malcom McLean was known for his meticulous attention to cost information. He tracked his costs very carefully, so he always knew how his trucking business was doing. This habit enabled him to make well-informed business decisions about the future. Looking back at those early days, his protégé, Paul F. Richardson, said, "I don't think anyone in this period had a better sense of north-south costs than Malcom. For ten years and more, he personally watched his trucks like a hawk. He knew every angle of costing. It is a fair question of whether he was thinking at that time about ships-versus-trucks: how much cheaper it might be to put trucks on ships. Certainly, he had the figures in front of him all of the time. Every truck that went out was an object lesson for him. He could do the figures for a run standing by the truck cab."[6]

The trucking business had always been subject to some federal regulation ever since the passage of the Motor Carrier Act of 1935. However, in the 1950s, as the trucking business matured, the individual states began adopting weight restrictions and levying fees. Truckers had to deal with a patchwork of state regulations, and they could be fined for excessively heavy loads. The problem became one of maximizing the load without incurring any fines. It was particularly problematic for truckers trying to pass through Virginia. The weight requirements in that state made it nearly impossible to make a profit on the load. R. Kenneth Johns, a longtime senior employee of McLean's, recalled, "If you were going through Virginia, you had to be underneath low weight limits. If you went over them, they'd get you for sure. Your company couldn't make money if you were under the limit, and wouldn't make much money if you were right on it."[7]

Why did the state of Virginia apply such severe weight requirements on truckers crossing the state? It was due to the political influence of two powerful railroads. There was considerable antagonism between truckers and railroaders. The Norfolk and Western, headquartered in Roanoke, dominated the east-to-west routes. Meanwhile, the Seaboard Airline, headquartered in Norfolk, dominated the north-to-south ones. "Those rail lines used their political influence to make life hell for truckers," said Johns. "At some point, Malcom figured 'life is too short. There's got to be a better way.' Virginia was right in the middle of his New York runs. His drivers had to cross it all of the time."[8]

McLean's first idea was simply to drive the trailers onboard a ship. You could roll the trailers on at the point of origin and then roll them off at the destination, a process known as "ro-ro." As he was mulling this concept over, he placed a telephone call to Brown Industries, a manufacturer of truck trailers in Spokane, Washington. He spoke with Keith Tantlinger, the vice president of engineering at the firm. It was a significant conversation. Tantlinger held a bachelor's degree in mechanical engineering from the University of California, Berkeley. During World War II, he had worked for Douglas Aircraft, designing warplanes.[9]

Malcom explained his idea about driving tractor trailers onboard a ship. Tantlinger said, "Why not just lift them onboard and leave the wheels behind?" McLean replied, "Can you do that?" Tantlinger said, "Sure, I can. I can even stack them, one on top of another."[10] Meanwhile, McLean's thoughts were percolating. There had to be a better way to transport cargo. He quickly embraced Tantlinger's ideas. He realized "that ships would be a cost effective way around shore side weight restrictions . . . no tire, no chassis repairs, no drivers, no fuel costs. . . . Just the trailer, free of its wheels. Free to be lifted unencumbered. And not just one trailer, or two of them, or five, or a dozen, but hundreds on one ship."[11]

These were the thoughts of a great innovator. McLean has been legitimately compared to Henry Ford and Thomas Edison. At this early stage, McLean was thinking of some sort of north-south shipping along the East Coast. To bring these thoughts to fruition, McLean benefitted from an important piece of legislation, little known outside of the maritime community. This would be the Jones Act, with the formal name of the Merchant Marine Act of 1920, designed to protect the American merchant marine from foreign competition. Still in effect to this very day, the act prohibits foreign-built, foreign-owned, or foreign-flag vessels from sailing in coastwise trade between any two U.S. ports. It also prohibits direct trade between any U.S. port and U.S. overseas territories. Ships protected under the Jones Act are not allowed to be serviced or maintained at foreign shipyards. In addition, ship crews of these vessels must be made up of U.S. citizens or legal residents. Thus, as McLean was thinking things over, one thing he did not have to worry about was foreign competition.[12]

The next step was to persuade Keith Tantlinger to move to Mobile, Alabama, and join forces with McLean's Pan-Atlantic Steamship Corporation. For the next few years, the two men worked hand in glove. McLean has often been called "the Father of Containerization," which is fair; but in many accounts, the contributions of Tantlinger have been overlooked. True, McLean was the financial force behind containerization, but Tantlinger was the technical genius. He did all the basic engineering—things like the corner fittings on the containers and how they engaged with other containers. He designed the removable truck chassis as well as the strength criteria for the containers, not to mention the way that the cranes engaged the containers.[13]

McLean was decisive. He purchased a pair of T-2 tankers. These oil tankers had been produced in large quantities during World War II. There were so many available that they could be purchased inexpensively. In 1955, he arranged to have them sent to Bethlehem Steel's shipyard in Baltimore, where they were fitted out with "spar decks," or an extra set of decks, or racks, above all the piping and pumps. The idea, of course, was to build a sort of platform upon which the truck trailers (without chassis or wheels) could be placed.[14]

The first vessel to complete the modifications was named the *Ideal X*. No one knows for sure how the name came about, but experts have speculated that "ideal" refers to the intermodal concept that was being launched and the "X" to the crossing of the land and sea modes of transport into a single service. Before the ship could be pressed into service, there were a couple of regulatory hurdles. In 1955, the ship underwent a series of sea trials. The U.S. Coast Guard wanted to ensure that the ship would be safe for her seafarers. Meanwhile, the American Bureau of Shipping wanted assurance that the ship's seakeeping and stability would not be harmed with the load of trailer bodies. The *Ideal X* passed all the necessary tests, and soon her sister ship was similarly outfitted and renamed the *Almena*. Both ships were ready to launch the new service in the spring of 1956.[15]

FIGURE 3 The very first container is loaded aboard the *Ideal X* at Port Newark, New Jersey, on April 26, 1956. The ship is dressed with a string of international maritime signal flags as a sign of celebration of the occasion. (Courtesy of the *Journal of Commerce*.)

Converting those two tankers into a couple of jury-rigged container ships was a bold move. In theory, the idea should have saved money. Studies had shown that 60 to 75 percent of the cost of sending things by means of break-bulk cargo took place at the dock. In the conventional practice of the day, a break-bulk ship might rack up $15,000 worth of stevedoring expenses on a single port visit. In theory, McLean could get the job done for a mere $1,600, not to mention reduced loss and pilferage, but to do this he would need specialized ships and customized ports.[16]

A lesser man might have given up at this point, but McLean had courage. He was an innovator who had confidence in his vision for the future. He decided that if old surplus tankers were not workable, he would turn to old surplus cargo ships. He looked to the C-2 cargo ships, many of which had built by the U.S. Maritime Commission during World War II. I asked naval architect Charles R. Cushing about McLean's pattern of adapting old ships for containerization rather than buying new ones. Cushing explained, "Whenever Malcom saw something floating, he was immediately trying to figure out

FIGURE 4 A large crane loads another container onboard the *Ideal X*. The containers are placed on steel racks above the pumping equipment. McLean's business later became known as Sea-Land Service, a company with ships that carried truck trailers full of cargo up and down the East Coast of the United States. (Courtesy of the *Journal of Commerce*.)

how many containers he could put on it." Cushing went on to explain that it was ingrained in McLean to use secondhand ships. He never thought in terms of new construction. To obtain a new ship, he would have had to go through a bidding process to find a builder and then to the banks to get financing. In the end, it would take two or three years to get that new ship. Instead, McLean was taking old cargo ships, and he could turn them into container ships in 120 days. The process was not only cheap but also fast.[17]

McLean turned to the naval architecture firm of George C. Sharp to convert six C-2s at a cost of $3.5 million each. (The letter "C" was for general cargo ships, and the number indicated the size of the ship. C-2 was relatively small compared with later ships such as C-3, C-4, and so forth.) The first one to be finished was the *Gateway City*, which was able to take containers stacked four high in the ship's holds and two high on the deck. In total, it could carry 226 fully loaded containers, much more than the 58-container capacity of the *Ideal X*. Arguably, the *Gateway City* was the very first true container ship.[18] The ship was equipped with special cranes, one forward and one aft, each capable of lifting 60,000 pounds, for the loading and unloading of the containers. The cranes could unload one container and bring another one onboard in about five minutes.[19]

Malcom McLean was a builder. He was always looking for ways to expand his business—whether it was acquiring new ships or developing new ports. All of this took a great deal of money, and this was money that he did not have. However, it all worked out because of an unlikely partnership with a financial backer—Walter Wriston of the National City Bank of New York (later Citibank). During the 1950s, Wriston was a young and ambitious banking executive at a time when the industry was rapidly changing. Up to this point, banking had been slow, stodgy, and conservative. Traditional bankers were assigned a slice of geographic territory. Their job was to cultivate clients in that territory, regardless of the industry. The focus was not on technical expertise but on socializing with the clients, usually with social drinking and golf.

Under the new system that was emerging, clients were organized not by region but by industry. Wriston was put in charge of the transportation industry—aviation, rail, shipping, and trucking. Wriston had no interest in schmoozing with clients. Instead, he was aggressively focused on the mission of finding suitable clients. He did his homework. In reviewing the bank's portfolio, Wriston realized that there was an overlooked opportunity in investing in trucking. As he studied lists of the top ten trucking companies in the United States, he kept coming across McLean Trucking. Following his hunch, Wriston made a trip to North Carolina.

On the surface, it would have appeared that the two men had little in common. McLean, the southern trucker, had finished high school in Winston-Salem. Wriston, the polished banker, had a bachelor's degree from Wesleyan University and a master's degree from Tufts. Notwithstanding this apparent educational and cultural gap, the two men instantly connected. Both were ambitious, impatient, aggressive, and focused on the job at hand. Wriston's biographer related a story that illustrates how that strong connection came about.

It seems that Wriston noticed that the sides of McLean's trailers were corrugated, not smooth. The highly educated Wriston said, "It seems you'd save gas if the sides were smooth, so you'd have less wind resistance." McLean politely replied, "You're right, except that I had a study made of the prevailing winds on the McLean routes that showed we pick up maybe a mile a gallon when we're pulling heavy freight with a following wind."[20] Wriston suddenly realized that he was dealing with a very smart person. Over the ensuing years, Wriston was willing to lend McLean the money to pursue his vision. Wriston was not afraid to make decisions. There was real risk in backing a new way of moving freight. Of course, it was not just out of friendship. Rather, Wriston was always focused on the success of the bank.[21]

The work of converting the first cargo ship into a container ship, to be named SS *Gateway City*, was carried out at a shipyard in Chickasaw, Alabama, on Mobile Bay, in 1957. By all accounts, the work went smoothly. Near the end of the process, a number of tests were carried out on both the ship and the cranes. The tests of all this new equipment were going well. In the final day of testing, while dignitaries were enjoying a fine lunch ashore, Malcom McLean went out to the dock to admire the ship. There he came across a man, whom he did not know, shaking his head in disgust. The man turned to Malcom and said, "You know what? We ought to sink the f-cker right now." It turned out that the man was an official of the International Longshoremen's Association. It was a foreboding warning of labor problems that lay ahead.[22]

It was a big day in early October 1957 when some 4,500 tons of cargo were placed onto the *Gateway City*, which left Port Newark for Miami and Houston with a full load. It was a conceptual breakthrough. At Port Newark, the ship was self-loading. At its destinations, it was self-unloading. In addition, at each end, the process took about one-sixth the time of handling conventional break-bulk cargo. We can only imagine the pride McLean must have experienced as the ship left, witnessed by dignitaries from the maritime industry, the military, and the government. Speaking at the event was Representative Herbert C. Bonner, Democrat from North Carolina, who was chair of the House Merchant Marine Committee. He called the container ship idea "the greatest advancement" in merchant marine history. The next day, a reporter from the *New York Times* described the event as "a silk-smooth operation that marked the first major change in coastal shipping in decades."[23]

Because of the successful sailing of the *Gateway City*, McLean was praised by the press as a maritime visionary. This well-deserved praise was not without irony because McLean did not come from a maritime background. Instead, he was a southern trucker with no sense of the romance of the sea. He had no childhood fascination with sea stories. For him, it was all business—a ship was like a tractor pulling hundreds of trailers. McLean saw himself as a hardheaded, practical executive. He told his associates that everything starts with a sale. He often said, "You sell, price, and move—in that order of importance." Without the sale, nothing happens. However, if you get the sale, but you do not price it right, that is no good. You have to convince the shipper that you are offering him a good deal. In terms of price, you want to make it enticing enough to agree to the deal; but, at the same time, you need to make a profit or you go out of business. Then, of course, you have to deliver the goods.[24]

In the same practical spirit, early on Malcom McLean decided to do away with the naming of ships in his fleet. He would just assign them numbers, such as Containership Number 1, and so forth. Fortunately, for the sake of nautical tradition, one of his captains came to him and said, "You know, no captain wants to go down with Containership Number 4."[25]

It is to McLean's credit that he respected the people who ran the ships—what they did in going to sea. At the same time, he saw his ships and their officers and crew as just one part of the whole picture of moving freight. He was in love with neither ships nor containers. He was often heard to say, "We don't move containers. We move freight."[26] Nonetheless, McLean readily agreed with the captain that the ships should have names. The first ship in the C-2 class was given the name SS *Gateway City*. The historical record is unclear on how that name was chosen, but we can readily speculate that Newark, New Jersey, was indeed that Gateway City. In any event, McLean seemed to begin enjoying the process, naming the remaining five C-2 converted cargo ships with names echoing his southern heritage.

First, let us consider the SS *Azalea City*, clearly named for Valdosta, Georgia. It seems that a parks supervisor in that city, Richard J. Drexel, took it upon himself to begin planting azaleas on public property throughout the city. Next, he began giving away azalea bushes to homeowners and churches. Drexel himself never pushed for the title "Azalea City." The title became official only in 1947, after being advocated by local garden clubs.[27]

Second, the next converted cargo ship given a southern name was SS *Bienville*, in honor of Jean-Baptiste Le Moyne, Sieur de Bienville, who was an early governor of French Louisiana. He is known historically for his many negotiations with American Indians and for his having selected the site for the present-day city of New Orleans. Today, Bienville Parish, a rural area in northern Louisiana, is named for him.[28]

The third converted cargo ship to bear a southern name was the SS *Raphael Semmes*, honoring the Confederate rear admiral known as the captain of the CSS *Alabama*, a Confederate commerce raider. The Confederate Navy was not strong enough to take on U.S. Navy directly, so Semmes focused on sinking merchant vessels flying the American flag. Here, he was very successful, sinking some eighty ships. Known for his bravery, he sustained the morale of southerners for a time. In the final duel between the CSS *Alabama* and the USS *Kearsarge* off the coast of France in June 1864, Semmes lost the battle and lost his ship, but that loss did nothing to damage his reputation as an honorable leader in the service of the lost cause.[29]

The fourth such ship was the SS *Beauregard*, named for Pierre Gustave Toutant Beauregard, prominent general of the Confederacy during the American Civil War. He was the first Confederate brigadier general, and he was in command of the defenses at Charleston, South Carolina, at the start of the Civil War at Fort Sumter in April 1861. Three months later, he won the First Battle of Bull Run near Manassas, Virginia. He is perhaps best known for his defense of the city of Petersburg, Virginia, from attacks by superior Union Army forces in June 1864. Following the war, he returned to Louisiana, where he became an advocate for Black civil rights.[30]

The fifth and last ship in this class was the SS *Fairland*, a traditional appellation for the American South. The Irish poet John Boyle O'Reilly, a contemporary of Oliver Wendell Holmes, captured that term in his poem "From That Fair Land and Drear Land in the South." In that nineteenth-century poem, he addresses the notion of southern hospitality with ambivalence. He refers to the South as both "fair" and "drear." Southerners typically welcomed friends and neighbors with open arms, but the poverty that followed the Civil War curtailed that hospitality, and the atmosphere was "dreary." O'Reilly was a critic of the myth of southern hospitality. Yet, for a son of the South like Malcom McLean, the notion of a "Fairland" was valid.[31]

Things were moving quickly. McLean had now switched from converted T-2 tankers to converted C-2 cargo ships. Along with this change, he upgraded the containers from 33 feet long to 35 feet, the maximum permitted on eastern highways. Fruehauf Trailer Company of Detroit, Michigan, built the new trailers. The 33-foot containers were phased out, as were the *Ideal X* and its sister ships. All of the early converted tankers, the pioneers in coastal container service, were unsentimentally sold by February 1958. The *Ideal X* was scrapped in Hirao, Japan, in 1965.[32]

In the very early days of Pan Atlantic's coastwise container shipping, McLean could depend on established customers who already relied on his north-south trucking company's routes. His first accounts included Nabisco, the multinational company with bakeries in the Northeast sending cookies and crackers to the South; Hublein Spirits, based in Hartford, Connecticut, known for premixed martini and Manhattan cocktails; and the Gillette Company, known for safety razors and shaving supplies. However, McLean needed to expand his sales force and find new customers. He began to hire young men, recent college graduates, many with athletic backgrounds. Their job was to persuade conservative and skeptical shippers to try the new mode of coastwise shipping.[33]

One of those new hires was R. Kenneth Johns, an Alabama native and a graduate of Auburn University, who joined the company in 1957. Johns believed in the mission of the company, and he became a successful sales representative, later becoming president of Sea-Land Service. McLean was a cheerleader for his sales force, and he inspired remarkable loyalty among all of his employees. Years later, Johns recalled, "Before one singe trailer had been shipped, Malcom had convinced all of the young guys he'd hired for his sales force that this thing was going to work. . . . And if he hadn't believed in what he was doing, we'd have been skinned when we made our pitches to potential customers." Selling coastal shipping to customers who were familiar with reliable over-the-road trucking was not an easy job. Most of these customers took a wait-and-see approach. They needed to be convinced. It took time to win them over. As Johns later explained, "When reports came back that trailers shipped by sea actually arrived safe, unbuttered, etcetera, the next sale was easier. There was at least a twenty percent saving for using us. . . . The real secret of containerization, for me, was as much in the selling as in as in the technology. The latter was a breakthrough, but you had to make people aware of it."[34]

It was hard to argue with McLean's concept. It kept working and drawing favorable attention from shippers. McLean's company, Pan-Atlantic Steamship Company, specialized in trade between Port Newark and southern ports such as Miami, Tampa, and New Orleans. Typically, southbound ships might carry carpet lining, tires, foodstuffs, bicycles, pipe, tubing, beer, cat food, applesauce, and potato chips. Northbound ships might carry citrus products, chemicals, plastics, rice, and sugar. Press coverage was generally favorable. Take, for example, the story of the loading and unloading of the SS *Raphael Semmes* at Port Newark in November 1958. The reporter told of a tractor-trailer pulling alongside the ship at 3:06 P.M. on a Friday afternoon with a van of twenty tons of packaged merchandise bound for the Miami area. The van was quickly loaded aboard the ship. At 3:11 P.M., the same truck drove away from the pier with southern products for local distribution. In other words, within five minutes, some forty tons of merchandise had been transferred between ship and truck. The whole process of loading and unloading the ship took thirteen hours, compared with the eighty-four hours it would have taken using conventional break-bulk methods, the loading of cargo individually in bags, boxes, crates, drums, or barrels.[35]

McLean's method saved not only time but also labor costs. The loading and unloading of the SS *Raphael Semmes* required only 42 dockworkers. To handle a similar ship with conventional break-bulk cargo would have taken 126 men. There was a tremendous cost savings here. The resulting job losses were controversial and very upsetting to the International Longshoremen's Association. Of course, much of the labor savings went into offsetting the cost of specialized loading and unloading equipment. Another problem was that carrying cargo in containers does not use the space aboard the ship in the most efficient manner. Some space is wasted. The company estimated that typically only 60 percent of the ship's cargo capacity was used. McLean explained, "We sacrifice tonnage for quick turn-around in port. That's the theory of the trailership. A ship earns money only when she's at sea. Costs are fixed at sea. Where costs rise is in port. The quicker you can get back to sea, the more money you keep."[36]

Some innovations spread fast, but containerization was slow to catch on in the early 1960s. Despite containerization's obvious advantages, the shipping industry moved cautiously. In 1960, according to the Maritime Administration in Washington, only 5 percent of the nation's oceangoing cargo fleet was capable of handling containers. Industry leaders were aware, of course, of the potential savings in insurance premiums due to the low rate of pilferage. Moreover, they knew of the savings to be had in time saved loading and unloading. On the other hand, they knew, as we have seen, that it would be difficult to get dockworkers to accept the new system. In addition, they worried about how to consolidate into a full container many small packages from diverse shippers. They also worried that full containers would be shipped overseas, only to be returned empty—a problem known as "deadheading." In late 1960, only 2.8 percent of the annual trade volume coming out of the Port of New York was in containers; most of it was still break-bulk.[37]

In the early 1960s, one key reason for the delay in implementation of containerization was the lack of standards. Each shipping company had its own individual standards resulting in a lack of interchangeability among rail, highway, and sea transport. In April 1961, the Federal Maritime Board took steps to bring order to the chaotic situation. Thomas E. Stakem, chair of that board, appointed by President John F. Kennedy, announced that the government would subsidize or finance only those containers that met the standard sizes authorized by the American Standards Association. In other words, approved containers would have a cross-section of 8 feet by 8 feet and standard lengths of 10, 20, 30, and 40 feet. "The approval of standard sizes is of great importance to the merchant marine," Stakem said. "The use of containers by ocean carriers as a means of reducing handling costs and providing improved shipper service is steadily growing."[38]

Not only did containers need to have standard sizes, they also needed to have uniformity in the fittings and attachment points for the stacking and locking of containers. This problem was highlighted in September 1961 at the International King's Point Cargo Handling Exposition, sponsored by the U.S. Merchant Marine Academy at Pier 9 on the Hudson River. The exposition revealed that, all too often, containers from different manufacturers were incompatible when it came to fixing them securely together. Organizers of the exposition called for the adoption of standardized fittings and hardware.[39]

As other shippers hesitated, Malcom McLean forged ahead. By the early 1960s, he retired the pioneering Pan-Atlantic name and logo. His company was now known as Sea-Land Service, a name that better described his assemblage of trucks, trailers, and ships. At the same time that the Port Authority was investing in container facilities, McLean was investing in ships. Up to this point, McLean had been using surplus World War II–era tankers and adapting them so that some containers could be placed topside. What he now needed was purpose-built container ships, designed as such from the keel up. However, that solution would have been too expensive, so he did the next best thing. He bought four T-3 tankers from Esso Standard Oil. The next step was both bold and imaginative: McLean had the shipyard workers cut off both the bow and the stern of each ship. The middle part, the tanker part, was discarded. It was replaced with a brand-new 417-foot middle section expressly designed to handle containers. This new midsection was built in Hamburg, West Germany, and towed across the Atlantic. It was then matched up with the old bow and stern. It was certainly not the ship you would have designed in the absence of financial constraints, but it worked.

The first of these new ships, formerly the SS *Esso New Orleans*, was finished up at the Todd Shipyard in Hoboken. It was christened the SS *Elizabethport* in honor of the new terminal complex. The ship set out from the port for which it was named on September 8, 1962, for Los Angeles and San Francisco. Several days later, the *Elizabethport* became the first all-container ship to pass through the Panama Canal. After eighteen days, on September 26, the ship passed under the Golden Gate Bridge and arrived in San Francisco, thus inaugurating container ship service between New Jersey and the West Coast.[40]

In an interview with Edward A. Morrow of the *New York Times*, Paul F. Richardson, vice president and assistant general manager of Sea-Land, said:

"We don't think of ourselves as primarily a ship operator. We are basically truckers. What we are doing is making use of the cost-saving ocean link between the highway systems of major manufacturing and marketing areas and combining this operation with the flexibility of inland trucking. In essence our new ships each represent a tractor which can pull 474 trailers over the ocean highway."[41]

Thus far, we have explored the birth of containerization under the leadership of Malcom McLean in terms of finding the right ships and developing the right boxes. We have touched on the importance of financing and salesmanship. Of equal importance was the parallel development of a specialized port set up to handle containers. That, of course, would be the Elizabeth Marine Terminal, or Port Elizabeth, as it is commonly known, the subject of our next chapter.

The Rapid Growth of Containerization

• • • • • • • • • • • • •

In the early days of containerization, Malcom McLean was pushing ahead on two fronts—ship development and port development. They were of equal importance. It was of no use to have container ships if they had no place to load and unload their cargo. At this time in the late 1950s and early 1960s, the shipping industry was still very conservative and resistant to change. McLean's agents would have to persuade reluctant port operators to accept containers. To make it easy on these port operators, McLean equipped his early vessels, the converted C-2 cargo ships, with shipboard cranes. This plan worked fine for a while. For instance, in the case of Port Houston, the container ship could offload its own cargo onto an empty parking area, surrounded by temporary buildings. It was not ideal, but it worked. One of the reasons that it worked was that the converted C-2 cargo ships carried only 226 containers, not so many as to overwhelm a small pier. In other words, in these very early days, McLean's ships did not have the luxury of dedicated container ports. They had to make do with makeshift solutions.

There were real problems with onboard cranes. For example, if a given crane was broken, the ship was unable to unload its cargo until the maintenance yard could make repairs. In addition, the cranes were exposed to salt air and were corroding. The company lost revenue because it was very hard to keep the cranes in reliable condition. On the other hand, the ships could go to new piers where there were no cranes ashore. Those piers were just parking lots; so, for the pioneering ships, it made sense to have onboard cranes. For an expanding company going into new trades where the local people did not know about containerization, the ships that had cranes were perfect. All you had to do was to arrive, and you could put the containers on the dock.[1]

Meanwhile, a very young naval architect within the Sea-Land organization was starting to have doubts about continuing to use shipboard cranes. At this time, Charles R. Cushing was one of the lowest men on the corporate totem pole. Throwing caution to the wind, Cushing sat down and wrote a long memo to Malcom McLean arguing for the use of shore cranes. As a result, he was nearly fired because in the distribution list of this long report was everyone down from McLean to Cushing's boss. Cushing got a call from his boss the next morning, "What the hell do you think you are doing?" The situation looked bad.

Then Cushing got a call from McLean, "Charlie, can you come up to my office? I want to talk to you about your memo." Cushing was worried. When Cushing got to McLean's office, McLean said, "It's a damn good idea. Yeah, we will keep some of the ships with cranes, but let's get going with shore cranes." Cushing was relieved. He was not in trouble after all. As the conversation continued, Cushing pointed out that they could put more containers on a ship if that ship did not have cranes onboard. Cushing had done his homework. He pointed out that if they replaced a crane with just two more containers, that would result in $1 million more in revenue over the life of the ship. In retrospect, this was an important encounter. There would be an increased focus within the Sea-Land organization on developing improved ports with their own cranes. Interestingly enough, there would be a place to put this idea into practice—Elizabeth, New Jersey.[2]

During this early period of slow adoption of containerization, an important development was taking place—the construction of the Elizabeth Marine Terminal just south of Port Newark. To understand how this development came about, we have to go back to 1955. At that time, everyone realized that freight tonnage at Port Newark was increasing dramatically. Port Authority maritime director Lyle King decided that more pier space was needed and that the new space should be developed to handle containerized cargo. He was especially receptive to the container concept from his experience with it while serving as a colonel in the Army Transportation Corps during World War II.[3]

Working behind the scenes, King approached New Jersey governor Robert Meyner, and he suggested that the governor ask for a study. It was a classic move from the Port Authority's playbook. The agency did not want to appear grasping. They would quietly get the governor to make the first move. Then they would graciously agree to the request for a study. Somehow, the study would show a real need that the agency could address. Sure enough, in December 1955, Governor Meyner announced that the Port Authority would expand into the city of Elizabeth and build a new port on 640 acres of tidal marshland.[4]

Construction began in 1958. In a bold and innovative move, the Port Authority decided from the very beginning that this new port would be designed as a container port. The plan called for a 9,000-foot channel, 800 feet wide and 35 feet deep, to be built at a cost of $150 million. There was to be thousands of feet of wharf frontage.[5] Interestingly enough, the new channel would be dug right along the political dividing line between Newark in Essex County and Elizabeth in Union County. Thus, the new channel would add to Port Newark on the northern side, and it would create Port Elizabeth on the southern side.

Dredging began on July 15, 1958. Captain Charles L. McLeod, a veteran of fifty-seven years in the dredging business and superintendent of the Atlantic, Gulf, and Pacific Company, supervised the work, under a $4,872,000 contract. The work was expected to take fourteen months of round-the-clock operation. McLeod deployed the suction dredge *Barlow* that was capable of rapidly gathering up bottom sediments of sand, gravel, muck, and mud, which were drawn through 8,000 feet of pipe and then deposited about a mile inland. Most of that material that was removed from the new channel was used as a foundation for the new buildings and storage areas.[6]

Having allowed time for the entire fill to settle and become firm, the next steps were announced by the Port Authority on July 14, 1960. They called for the construction of three ship berths along the north side of the new channel in Port Newark. These were designated locations where vessels could be moored for loading and unloading, to be completed by the summer of 1961, at a cost of $3,645,000. These three berths were to be the first of eleven new ones at Port Newark. Added to the preexisting twenty-eight berths at Port Newark, it would give Port Newark a potential capacity of some thirty-nine berths. At the same time, the Port Authority awarded a contract of $2,197,600 for the construction of four new berthing facilities on the south side of the channel for Port Elizabeth.[7]

On March 22, 1962, Governor Richard J. Hughes announced that the Port Authority had signed a twenty-year lease of its new facilities at Port Elizabeth with McLean's Sea-Land Service. The announcement took place, with much fanfare, at a luncheon meeting at the Military Park Hotel in Newark. Some eight hundred guests attended the event arranged by the Newark Association of Commerce and Industry and the New Jersey Citizens Highway Committee.[8] With the lease in hand, Malcom McLean turned to a young employee, Ron Katims, a recent civil engineering graduate of Cornell University, and gave him the task of designing the new container port. It was a daunting task because there was no template, no precedent, and no prior design for such a thing. The Port Authority had given Sea-Land a piece of land that had been brought up to grade but was otherwise undeveloped.

Years later, Ron Katims explained to me how the process worked: "The Port Authority gave us the land up to grade, and we had to provide everything above that—the pavement, the lighting, the electrical distribution, and the buildings. We gave the bill to the Port Authority, and then they figured the cost into the lease, based on a twenty-year payback." Katims went on to explain that the arrangement was just what Sea-Land needed because there was no immediate outlay of cash. However, there was a significant catch. The Port Authority held Sea-Land to a strict limit of $11 million, including the cost of engineering.[9]

I asked Katims how he tackled the problem of designing a container port from scratch, with no precedent, no template. "I started with a piece of paper. I then sat down with the operating people. We went through how we were going to operate. They told me how they wanted to operate, and I tried to translate that into a layout. We went back and forth with various layouts."[10] There were many planning meetings and discussions; but, in the end, a layout and a plan were developed based on the technology of the time.

I asked him to give me an example of the process. "Well, consider the rows of containers. A truck has to come in with a tractor and pull a container out. How much space should be between the rows? There was no way to know, so I went out there with a couple of trucks and drivers with some tapes and measurements. We figured out that the distance between the rows should be 60 feet. That was the turning radius needed. We experimented with actual trucks. There were no computer simulations at that time to do that kind of testing. There was a lot of trial and error, and we developed a plan." Katims went on to explain that he could have cut the distance down from 60 feet, but it would have created problems. If a truck driver had to wiggle and work, back and forth, then he would be wasting valuable time in delivering the next container to the ship. You have to have enough room to pick up the container efficiently, and yet you do not want to give them too much room. It was all decided by trial and error.[11]

I asked Katims about the advantages and disadvantages of grounded versus chassis container operations. On the one hand, a container with a chassis is a wheeled structure that carries marine containers for the purpose of truck movement. It has a steel frame, tires, brakes, and lights. On the other hand, containers can be grounded, without chassis, and they can be stacked, one on top of another. A key advantage of the grounded system is higher storage density.[12]

Katims told me that McLean strongly preferred a chassis operation: "Malcom McLean was a trucker. The minute a container came off a ship, it went on a chassis; and it became a truck." He did not like any grounding systems. His preference for a chassis system had implications for the pavement. If you were to put a heavy container directly on the ground, you have to have very strong pavement. With a chassis, the weight is distributed over several rubber tires, with less stress on the pavement, so you can use less expensive asphalt, with one small exception. At Port Elizabeth, Katims had to install concrete landing strips for the dolly wheels of the chassis. When a tractor offloads a chassis, the chassis has real rubber tires at the rear, but much smaller dolly wheels at the front. These small wheels would have punched right through conventional asphalt paving. Instead, the dolly wheels rested on concrete landing pads that were three feet wide to accommodate a variety of settings.

For piling up stacked containers on the ground, you need much stronger pavement, normally highly durable concrete. In the early days of Port Elizabeth, the containers were never stacked. Instead, they were placed side by side on chassis. While this system did not require strong pavement, it did require more land. An advantage of McLean's system was that there was less delay in picking up a load, which was better for the customer. When a trucker came into the port, the container was already on a chassis, and he took it out.[13]

I asked Ron Katims about lighting at Port Elizabeth. He explained that good lighting was important for both safety and security, especially since the port was envisioned to operate twenty-four hours a day. "It was a matter of testing. We went out there with a light meter to see what was needed for the drivers to read the numbers on the trailers. It was not like today when the driver uses an electronic scanner. We experimented, and we found that we needed a three-foot candlepower (a measure of luminous intensity) at the rear of the trailer. We placed light poles between the rows of backed-up containers."[14]

With the marshalling yard laid out, the next step was placing the buildings. The original plan called for a general office building and a marine operations building. It also called for a truck operations building, a truck maintenance garage, a general warehouse, and a building for the longshoremen.[15] Katims explained that he laid out the plan where the buildings should go, but that the architectural design work was outsourced to the architectural firm of Frank Grad and Sons of Newark. That firm was well known for many buildings in northern New Jersey, but the one that most impressed the Sea-Land team was the Prudential Tower in downtown Newark. Katims worked with Frank Grad to make sure that the buildings would please Malcom McLean, who wanted something that looked good, something of which he could be proud. Katims recalled McLean saying, "You know, when people walk up here, they want will want to work for us. If they see a sloppy, rundown warehouse, they will walk away; but, if they see a beautiful building, they will stay."[16]

The first building to be designed was for truck operations, a two-story building, with canopied trucking lanes off to the side, furnished with electronic weighing scales and pneumatic tubes to send documents into the offices. Katims knew that, in time, such paper documents would be replaced by computers, but he had to deal with existing technology. The idea behind the truck operations building was simple. A truck driver with a container to be shipped to another port would drive up. The operations people would tell him exactly where to drop that container, and then they would tell him exactly where to go to pick up a container that needed to be delivered locally. These tasks at Port Elizabeth were made simpler for Sea-Land because the firm was dealing with domestic Jones Act cargo, thus they did not have to deal with the complexities of customs. On the way out of the port, the truck would go through another building, the maintenance facility. There, Sea-Land would check on the road ability of the tractor-trailer—tires, brakes, headlights, and so forth.[17]

How to keep track of the containers? They used magnetized wooden blocks put up on a giant metallic bulletin board. Office workers could move the blocks from ship to land. It was like a primitive computer, with people moving the blocks around. In time, of course, they went from wooden blocks to IBM punch cards, and later, naturally, to computers. However, in the early days, there were occasional problems. Katims told of one such incident: "On a Saturday morning, one of the men who worked there brought his kids into the building. While he was doing his work, the kids took all the blocks down, to play with them."[18]

Comparable to the truck operations building was the marine operations building. Here the port engineer would work to ensure smooth port operations and to avoid congestion. He would facilitate the arrival and departure of the vessels. He would supervise the loading, unloading, storage, and

FIGURE 5 Ships unload at Port Elizabeth, New Jersey, on February 12, 2005. The port has a channel that is 9,000 feet long and 800 feet wide, allowing ships to be serviced on both sides of the channel. (Photo credit: Alamy Stock Photo.)

distribution of cargo. Most importantly, he would try to ensure quick ship turnover in port, since the ships were only making money when they were moving cargo. Nearby was a smaller building with maintenance supplies such as lubricants, paints, and welding supplies as well as replacement parts such as pumps, propellers, and shaft seals. From McLean's point of view, the building was most important for its third floor with a private dining room where he could entertain customers and investors. There was an outdoor balcony where he could impress his guests with striking views of the harbor.[19]

The largest building in this new port complex was the general cargo terminal. It had a clear-span interior 100 feet wide by 1,100 feet long, with no central columns. The center was generally kept clear, but along the sides were stacks of pallets and rows of forklifts. A two-story office building at its center broke its long horizontal shape. Along the two sides of the building were 170 truck-loading docks. Each dock had adjustable dock boards to accommodate different tailgate heights.[20] Most of the incoming freight was in less-than-trailer loads because most of the shippers were sending small shipments. The real purpose of the general cargo terminal was to *consolidate* these small shipments and load them into intermodal containers to take advantage of the ensuing cost savings. In these early days, most of the outgoing cargo was headed for Puerto Rico, under the protection of the Jones Act.[21]

The completion of this initial phase of construction of Port Elizabeth represented a real paradigm shift. Here was a brand-new port, costing millions of dollars, that was entirely dedicated to handling containers. Everyone in the industry was watching. Success was not a foregone conclusion. McLean knew that, for his experiment to work, there would have to be more ports dedicated to containerization. He turned to the young civil engineer, Ron Katims, to make that happen. Developing new ports was an interesting challenge. First, the people in the new location would have to be sold on the idea. Once they agreed in principle, Katims had to deal with the particulars of each different location. In these early days, in the late 1960s, McLean focused on ports under the American flag, protected from foreign competition by the Jones Act. Let us see how this worked, starting with the case of Puerto Rico.

It can be said that port development in Puerto Rico represented the true coming to life of container shipping. True, containers had been used in the coastwise service to Florida and Texas, but here was the first application offshore. Here the competition was not with rail or truck, but with other shipping companies, which were slow to adapt. Thus, Sea-Land quickly became the dominant carrier, and the company was straining to keep up with the growth of operations.

In those early days, the makeshift container operations were being run out of a hangar at the Isla Grande Airport, just across a small channel from Old San Juan. The ships were being docked on one side of the runway, but the marshalling yards for the containers were way over on the other side of the runway. Each container had to be transported by truck for a two-mile run. The loose freight from the containers was emptied onto the floor of a hangar. The truckers would then come and sort through the mess to find their cargoes for delivery. There were also fully loaded containers to be picked up by still other truckers. The whole arrangement was inefficient.

Malcom McLean was aware of the shortcomings of the Isla Grande facility, and he knew that something had to be done. Fortunately, he had developed a friendship with Theodoro Moscoso, who was an influential Puerto Rican business executive and politician. Moscoso was the head of a program known as "Operation Bootstrap." The idea behind this program was that Puerto Rico had to move beyond dependence on agriculture and to move more into industrialization. It seems that one day McLean and Moscoso were discussing their mutual interests over lunch at the Bankers Club in Old San Juan. McLean mentioned that there was a need for a new and better seaport in San Juan. Moscoso looked out the window and pointed to the city dump.

It was easy to see because it was always on fire. The two men agreed that it was a good spot for a new port.[22]

Soon, the Puerto Rican Ports Authority (PRPA) began constructing new berths in that area, and Sea-Land was negotiating for the use of those facilities upon their completion. At this point, Ron Katims was sent to Puerto Rico to participate in those negotiations and to make the area into a dedicated container facility. The PRPA built the berths and brought the land up to grade. Then, Sea-Land took the project from that point, just as they had done in Port Elizabeth, but with one big difference. Sea-Land developed the port, but with their own money. There was no reimbursement from the PRPA. As a result, there was a very strict budget. Fortunately, San Juan was a natural port, though with depth problems. Sea-Land had it dredged as far as possible.

Katims was originally responsible for the construction of three different improved ports in Puerto Rico for Sea-Land. The main port, and the easiest to reach on the north side of the island, was at San Juan. There was a secondary port at Ponce on the south side of the island, and another at Mayaguez on the west side of the island. However, it soon became obvious that it was better and cheaper to take the goods to San Juan, and then distribute the goods by truck to the rest of Puerto Rico. The local people were the ones who demanded the change.

For example, the manager of a supermarket located in Ponce insisted that his container should be offloaded in San Juan. Then they could run a truck across the island long before a ship could steam around to Ponce. It was not commercially feasible to have a container ship call on a port in Ponce because that ship would have to circumnavigate the island. Specifically, if a ship were to leave Jacksonville, Florida, on a Friday night, it would arrive in San Juan at seven o'clock the next morning. That supermarket in Ponce wanted the goods to be delivered by truck no later than Monday afternoon; but if the ship had to run around the island, the goods would not be there until Tuesday or Wednesday. In the end, Sea-Land could not make the port in Ponce work. Indeed, the situation in Mayaguez was even worse because a ship had to travel even farther. Over the years, as the roads in Puerto Rico improved, truck transportation was the obvious way to go.[23]

Throughout the 1960s, Sea-Land continued to expand its container business through the construction of new container ports. Each case was different, but the expansion followed a three-step pattern. The first step was an exploratory discussion. Would the prospective port authority be interested in the idea? Some ports were very conservative and resistant to change. Others were more nimble, adaptive, and receptive to the idea. The second step was an extended negotiation. Who would pay for the dredging and the grading? Who would pay for the construction of the facilities? What would be the terms of the lease? The third step was the most straightforward—the building of the port. Let us take up a few examples.

While Ron Katims was busy overseeing the construction of the port facilities in Puerto Rico, he was simultaneously planning to run ships from Sea-Land's new terminal in Port Elizabeth to the West Coast by way of the Panama Canal. He began negotiating with two ports on the West Coast—Oakland and Long Beach. Neither one was the obvious choice. The Port of Oakland, located within San Francisco Bay, adopted containerization before the Port of San Francisco. The Port of Long Beach, located about twenty-five miles south of Los Angeles, adopted containerization before the Port of Los Angeles. Both of these ports had managers who were visionary, farsighted, and willing to take a chance on the new technology.

The negotiations for the terminals in Oakland and Long Beach were challenging because Sea-Land wanted a change in the way that port charges would be assessed. At that time, all of the West Coast ports were charging cargoes coming off vessels a charge per ton, or wharfage. Instead, Sea-Land wanted to lease the properties at a fixed yearly rent, a more suitable and simplified accounting method to accommodate containerization. Both Oakland and Long Beach agreed to the new arrangement, but the other ports launched a number of lawsuits. There were hearings in front of the Federal Maritime Commission. Ron Katims had been trained as civil engineer, not as an attorney, but he was thrown into the conflict. Years later he wrote, "Since I was the chief negotiator for Sea-Land, I had to testify several times, and thus had my first-ever cross examination from some very tough lawyers." Fortunately, Sea-Land prevailed and the suits were dismissed. The terminals that were constructed in both ports were smaller versions of what had already been built in Elizabeth, New Jersey. Both projects made their deadlines and their budgets on a regular basis.[24]

The early 1960s were exciting times at Sea-Land. Containerization was catching the attention of the maritime industry. With the successful introduction of the concept at both Oakland and Long Beach, the logical next step was Alaska. Ambitious perhaps, but still protected from foreign competition by the Jones Act. In the winter of 1963, Malcom McLean directed Katims to make a trip to Alaska to evaluate Alaska Freight Lines (AFL), a company that he was considering to buy. AFL operated a system of tugs and barges that serviced the ports of Whittier and Seward. However, AFL was desirable because it had facilities in Anchorage, Alaska's most populous city, located in the south-central portion of Alaska, at the end of the Cook Inlet. That inlet stretched for 180 miles from the Gulf of Alaska up to Anchorage.

Katims and his team of evaluators reached the conclusion that Sea-Land could have its container ships navigate the Cook Inlet all the way to Anchorage, something that had not been tried before. AFL and other shippers had not attempted direct deliveries to Anchorage because Cook Inlet is not ice-free in the winter. However, the Sea-Land evaluators discovered that the inlet rises and falls with a thirty-six-foot tide, so the water is always moving in or out. Thus, this water movement breaks up the ice and allows for the possibility of ship movement in the winter. The Sea-Land group met with officials of the Anchorage Port Authority, who were interested in leasing their facilities. Sea-Land negotiated for a preferential usage agreement and a lease of land for container parking. Rather quickly, an agreement was reached. The Sea-Land group then flew to Seattle to negotiate a berthing arrangement with the Seattle Port Authority, which was rapidly concluded.[25]

Things looked good for Sea-Land that winter, but then a few months later something terrible happened. On Good Friday, March 27, 1964, as many Alaskans were sitting down to dinner, a great earthquake of magnitude 9.2 struck the Prince William Sound region of Alaska. The earthquake rupture started about seventy-five miles east of Anchorage. It lasted about four and a half minutes, and it is the most powerful recorded earthquake in U.S. history.[26] The city of Anchorage was hard hit, with buildings collapsing everywhere. Water, sewer, and gas lines were ruptured. As the tremors ended, tsunamis followed, demolishing the ports of Whittier and Seward.

On Easter Sunday, Malcom McLean asked Ron Katims for a report on the situation in Alaska. Katims told him what little he knew. McLean asked who was looking after Sea-Land's interests in Alaska. The answer was no one, so Katims was sent to Alaska on Monday to see what could be done. When he arrived in Anchorage, the city was in chaos. The National Guard had been called out, and many people had been displaced. He had trouble finding a place to stay, and he finally got a bed in a hotel ballroom that had been turned into a dormitory. His first meeting was with the mayor, who was understandably preoccupied with the disaster at hand. The mayor told Katims that the whole deal was off the table. Katims disagreed, and they proceeded to negotiate. Katims quickly offered to bring up two ships with emergency relief supplies. This goodwill gesture showed the mayor Sea-Land's humanitarian concern for the people of Anchorage. It saved the deal. In time, the new terminal in Anchorage was completed.[27]

Throughout the early 1960s, as we have seen, the Sea-Land executives, working from their headquarters in Port Elizabeth, enjoyed considerable success in pushing for container ports in Puerto Rico, Alaska, and the West Coast. To be sure, there were challenges in setting up these ports. However, these challenges paled in comparison with next set of challenges, halfway around the world, in Vietnam. As American involvement increased throughout 1965 and 1966, it became clear that the U.S. Army was having great difficulty in providing adequate logistical support for our troops. Early on, the U.S. military decided that the main supply route would be through a harbor in Cam Ranh Bay, an inlet of the South China Sea, some 180 miles northeast of Saigon. It was a natural, deep-draft harbor, surrounded by mountains and relatively easy to defend. Most of the cargo would arrive here, and then be distributed by smaller vessels up and down the coast as needed.[28]

Though Cam Ranh Bay was a natural harbor and an obvious choice for receiving supplies, there was a big problem. It had only one deep draft, stone finger pier, consisting of two berths that could accommodate two ships, very inadequate to deal with all the incoming supplies. Ships had to wait for weeks and sometimes months to unload their rations, rifles, ammunition, trucks, tires, plasma, and hundreds of other items vital to U.S. forces in Vietnam. It was clear that the army had to build new piers, but there was a major problem. There was a lack of local port construction materials such as structural timbers, creosoted timbers, and steel sheet piling. Even if those materials had been available, it would have taken eighteen to thirty-six months, with the labor of some four hundred workers, to build one additional permanent pier. It was simply too long to wait. Fortunately, the Army Corps of Engineers had a solution to the problem. They had similar problems back in World War II. Back then, Colonel Leon B. DeLong designed prefabricated floating platforms, or barges that could be put into position, jacked up, and turned into piers. Such barges had mobility. They could be towed to anywhere in the world. Once in position, they could be stabilized to resist wind, waves, and weather.[29]

Fortunately, there was a prefabricated DeLong pier, built in 1952, in storage in Charleston, South Carolina. By June 1965, it was refurbished and ready for service. The pier left Charleston for Vietnam on August 11, 1965. It was towed some 12,000 miles, and it arrived eighty-one days later on October 30, 1965, at Cam Ranh Bay. It was positioned at a causeway that had been provided for it. Fixing it into position required long hours of work in very hot weather, but by mid-December of 1965 the pier was operational. It was capable of discharging break-bulk cargo at the rate of 800 tons per day. Now, there were two piers—the old stone pier and the first DeLong pier. Combined, they could unload four ships simultaneously. It was certainly an improvement, but not enough to break the ship backlog of the many ships at anchorage.[30]

The Army quickly contracted with the DeLong Corporation for additional piers. DeLong, in turn, subcontracted the work to the Japanese at the Sasebo Dock Yard. It took the Japanese only ninety days to make a single DeLong barge. The barge was then towed to South Vietnam. The trip took two to three weeks, depending on the weather. Once a barge arrived at Cam Ranh Bay, it took twelve men about four weeks to secure the barge and turn it into a pier. By January 1967, there were four DeLong piers operating in Cam Ranh Bay.[31]

These DeLong barges were 300 feet long, 90 feet wide, and 13 feet deep. They could be towed where needed by tugboats. What made them unique were twenty-two round openings, or holes, through the deck where caissons, or very large pipes, could be erected and driven into the seabed. The caissons were 140 feet long and 6 feet in diameter and fabricated from 1.5-inch-thick steel. Each one weighed 79 short tons. Once the caissons were seated and could not be driven any farther, using hydraulic jacks the barges were raised up on the caissons and out of the water, to a desired height. Then, the excess pipe lengths could be cut off and removed. Welders would smooth the jagged edges, since each stub had to be capped off. The crew would put the cap into position so that the bolt holes would line up. Welders then would seal the cut-off stub to the cap acceptance collar. Finally, fenders to absorb the impact of docking vessels were bolted to each caisson and to the barge. Thus, in very short order, the barge had become a pier.[32] To be fair, the Army Materiel Command deserves credit for its pioneering effort to install DeLong piers at Cam Ranh Bay in the 1966–1967 period. It was a clever solution to the absence of docking facilities to unload cargo. However, this effort was insufficient to clear the backlog of ships waiting to be unloaded. Meanwhile, behind the scenes, back in the United States, Malcom McLean was aware of the deepening crisis in providing ships to keep up with sending supplies to Vietnam. McLean approached the Army with a solution. He went to the Pentagon without an appointment and waited outside the office of General Frank S. Besson Jr., commander of the Army Materiel Command. The general promised him half an hour. Once they got talking, the visit turned into several hours.

McLean started by explaining his revolutionary concept of ocean shipments by sealed and locked containers. To his credit, Besson listened patiently, though he had good reason to be hesitant. At this time, containerization was still an infant and untested technology. Many were dismissive of Sea-Land. They saw it as simply a niche carrier protected by the Jones Act and domestic cabotage laws. Still, the general kept an open mind. He authorized Sea-Land officials to go to Vietnam, to observe, and to come back with a report and recommendations. The Sea-Land people were to talk to commanders in the field and to get their thoughts. The cost of the trip would be borne by Sea-Land. They had to make their own hotel and other arrangements, except for transportation within Vietnam.

By the end of 1966, McLean had convinced the military in Washington that containerization was a solution to the logistical mess and glut of vessels attempting to supply the troops. In 1967, General Besson told McLean that, to implement containerization in Vietnam, he would have to get agreement from in-country commanders that it was wanted, needed, and could be handled. The general authorized a visit to Vietnam by Sea-Land officials to make this happen. Ron Katims, and the other Sea-Land executives, met with Major General Charles William Eifler, head of the Army's First Logistics Command. Eifler told them to go see the depots, talk to the marine terminal operators, and have them send him messages on their views of containerization.

Convincing skeptical engineers and port operators to accept containerization by placing container cranes on DeLong piers in Cam Ranh Bay was a difficult and lengthy process. There was considerable resistance to this new idea. By early 1968, Cam Ranh Bay now had six DeLong units, making 3,600 feet of berthing space. Even with this new berthing area, many vessels were still at anchor outside the harbor, waiting their turn to offload. The Sea-Land people had a plan to deal with the problem. They suggested having a vessel arrive every Wednesday and use one 600-foot berth reserved on a preferential basis. Containers would then be distributed up and down the coast by smaller feeder vessels equipped with onboard cranes. The port operators were skeptical, but the Sea-Land people pointed out that, in a reserved one-day period, they could discharge more cargo than all the other berths combined could discharge in an entire week. It looked like approval was just around the corner.

Later, there was meeting in Washington to wrap things up and come to an agreement. The meeting started out amicably, but all of a sudden General Eifler raised an unanticipated objection. Eifler said that, in his opinion, the strong winds at Cam Rahn, combined with the tides, would create a sympathetic vibration in the tall cranes that would bring them down. The general was worried that, if such an accident were to happen, he would lose the use of a pier at Cam Rahn. It was a powerful objection. Meanwhile, the consequences of such an accident would be disastrous for Sea-Land, which would lose its investment and possibly face bankruptcy.

The conversation was heated. Ron Katims was angry because he believed that the general's argument was baseless. Katims finally said, "General, you stick to being a general, and I will stick to being the engineer. What you have stated cannot and will not happen." The general was not accustomed to such straight talk, and he backed down. Katims had taken a big risk, but he got away with it.[33]

Sea-Land had come a long way. They had met with the logistical people in Vietnam. They had patiently explained their plan. They had answered all the objections. They had persuaded the top brass at the Pentagon. However, they were dealing with the government, so they could not just sit down and negotiate a contract. They would have to submit a competitive bid. At this time, most contractors in Vietnam were submitting cost-plus bids. Such a bid would reduce the risk of a loss for the contractor. Almost no one was submitting fixed-price bids for work in Vietnam. Doing so was seen as too risky. However, McLean went ahead and submitted a fixed-price bid. Why did he do this? McLean himself never explained his decision, but I can hazard a guess. First, McLean was known to be very good at costing out a project. He was careful with the details. Second, he knew that the government would be pleased by a fixed-price bid. For the government, there would be no cost overruns, no unexpected surprises. Third, his fixed-price bid would frighten off potential competitors, who would not want to take such a risk. In any event, the strategy worked. Sea-Land signed a lump-sum contract with the Military Sealift Transportation Service.

With the contract in hand, Sea-Land could begin the actual work of installing the first-ever container pier in Vietnam. The plan was to take a regular DeLong pier and install two specialized, high-speed, dock-mounted cranes. The problem was that the DeLong barge required stiffening of the internal supports so they would not buckle under the weight of the cranes that each weighed a million pounds. In addition to the crane problems, Ron Katims had to arrange for parking space for the containers and power on shore for the refrigerated units. All of this had to be accomplished within six months in order to accept the arrival of the first vessel without cranes at Cam Rahn. Katims wrote, "This was more than a crash schedule. It was a race to the finish or lose the entire game."[34]

Sea-Land decided to gather everything that was needed in Manila, Philippines, where it would be loaded onto two barges and towed into Ram Ranh Bay. The barges had onboard two cranes, the equipment to erect the cranes, trailers for housing the workers, food, electric generators, the pieces to strengthen the barges, and crane rails. They tried to think of everything because they knew that they could not count on any help from the military. It worked out well because they had so many supplies that were able to barter some of them with the military for meat, eggs, and fresh food. As the items arrived in Manila; they were placed on the barges that had armed guards because piracy was a big issue in the Philippines.

At the same time, a subcontractor was found to carry out the work to stiffen the DeLong pier. The front man for this contractor was an Australian named Alan Keats, whose scheme was to fit out a barge with living quarters for Singapore laborers he would bring to Vietnam to do the work. The barge would tie up next to the DeLong pier. Finally, by early April 1968, the two barges from Manila and the single barge from Singapore arrived within hours of each other. Now the work could start in earnest. Huge holes were cut in the sides of the DeLong pier to supply air for the welders. It was difficult work because the temperature inside the barge was 125 degrees. Small pieces of steel had to be welded to the structure of the barge. Most of the work had to be done on ladders. The men worked in thirty-minute shifts, the most that they could stand.

Once the barge was strengthened, the cranes could be installed. The work proceeded without incident. When the first crane was completed, they put up a red warning light at the uppermost tip, some 135 feet up in the air. Years later, Ron Katims recalled, "That night from my sleeping bag in the trailer I could see the light very clearly. Then the wind picked up to forty knots, and all I could think of was General Eifler's theory that the cranes would be affected by the tides and winds and would not survive. I never slept a wink and quietly in my mind tried to calculate the amount of motion that the one standing leg was experiencing and would that be enough to have the structure fail." Fortunately, it did not fail, and those cranes worked and stayed in Vietnam for many years.[35]

Selling the concept of containerization to the U.S. Army during the Vietnam conflict was a remarkable accomplishment for Malcom McLean and his company, Sea-Land. We have seen how this U.S. shipping line undertook, as a private carrier, the job of supplying the armed forces. Clearly, McLean took huge financial risks in doing all this under a lump-sum contract. There were plenty of obstacles and frustrations along the way. Where did McLean find the courage to do this? We may never know, but I would suggest two forces were at work. First, McLean honestly and altruistically believed that he was doing his patriotic duty to help the U.S. Army solve a terrible logistics problem. Second, on a more practical level, McLean figured that if he could sell the containerization concept to such a huge and hidebound organization, he could sell it to anybody.

In any event, by the end of the 1960s, Sea-Land, based in Port Elizabeth, had the world's largest fleet of container ships, with over 18,000 containers in service. The company was serving twenty-two ports in the United States, Puerto Rico, the Virgin Islands, the Dominican Republic, Okinawa, and Europe.[36] The rest of the world was watching and quickly catching on to the new idea. Today, we find containers everywhere.

5

From the Ocean to the Docks

●●●●●●●●●●●●●●

The first European to sail into what is now New York Harbor was Giovanni da Verrazzano, who found a safe harbor there in 1524. What he saw was an unspoiled virgin land. Years later, the English navigator Henry Hudson explored the same harbor, and he quickly realized that it provided easy access to the interior. Now nearly five hundred years after Verrazzano's discovery, this region has become densely populated with a highly developed economy.[1]

The harbor worked well throughout the colonial era and the nineteenth century. It offered a protected anchorage, and it was deep enough to handle most ships. However, at the dawn of the twentieth century, there was a realization that, as ships got bigger, with deeper drafts, something had to be done. The traditional approach to the harbor was tricky and complex, as ships navigated to avoid shallow areas. It became clear that what was needed was a straight and deep channel leading from the ocean to the docks, but such a project would require political consensus and considerable funding. Fortunately, there was a man of vision with enough skill and influence to push for the project. That man was John Wolfe Ambrose.

Ambrose was an Irish immigrant who worked his way through New York University and Princeton. Educated as a civil engineer, he established his own contracting business in New York City. Once established as a contractor, he turned his attention to fighting for deeper shipping channels. He was civic minded and knew that the future of the city depended on maritime commerce. He met with leaders in the U.S. Senate in 1899 and persuaded them to appropriate the money for dredging the channels, thus enabling the city's economy to prosper and to compete in international commerce. The Ambrose Channel and Light were both named for him.[2]

When the first phase of the Ambrose Channel was completed in 1907, there was a tremendous wave of enthusiasm in the maritime community. The new channel went in a straight line, eliminating many twists and turns. It was both safer and six miles shorter. The Army Corps of Engineers had successfully dredged it to a width of 800 feet and a depth of 32 feet at low water. They were already planning to expand it to a width of 2,000 feet and a depth of 40 feet. Everyone was excited when the Cunard Liner RMS *Caronia*, a large transatlantic steamer, passed safely through the new channel.[3]

Historically, a lightship had always marked the approach to New York Harbor. Such a ship took the place of an offshore lighthouse because the water was too deep. Along with the new channel, there was to be a new lightship to mark the entrance to the harbor. In early December 1908, the new *Ambrose Light Vessel* (LV-87) replaced the old *Sandy Hook Light Vessel* (LV-51).[4] The average life of a lightship on station was about twenty years. During the twentieth century, there were three different lightships posted at the Ambrose station.

First was the *Ambrose* (LV-87), which served from 1908 to 1932. It was the last of the steam-powered vessels to hold that post. The *Ambrose* (LV-111) that served from 1932 to 1952 replaced it. This ship was diesel powered, and it served throughout World War II. It was equipped with radar in 1945. The last lightship to mark the Ambrose Channel was the *Ambrose* (WLV-613), which served from 1952 to 1967.[5]

The third and last lightship marking the Ambrose Channel was replaced in August 1967 with a fixed structure of the "Texas Tower" type. It was a platform supported by four legs placed into bedrock on the ocean floor. Atop the platform was a light tower and a deck for helicopter landings. Because of its fixed location, it was an improvement over the previous lightships, attached by cables to buoys, allowing the ships to move with the wind and the tides.[6] Built at a cost of $2.4 million, the Ambrose Offshore Light Station, as it was called, was equipped with a powerful light beam of six million candlepower, which could be seen for twenty-one miles. Another advantage was that the new tower required a crew of only six men, whereas the lightship had required a crew of sixteen men.[7]

Building the light tower seemed like a good idea at the time. Indeed, it worked well for some twenty-nine years, but then in October 1996, the Greek

FIGURE 6 On April 22, 2010, tourists flock to Pier 17 at the South Street Seaport and to Pier 16 to see the Lightship *Ambrose* (LV-87) that marked the Ambrose Channel from 1908 to 1932. (Photo credit: Terese Kreuzer / Alamy Stock Photo.)

oil tanker M/T *Aegeo* hit the tower, causing considerable damage. Repairs were made; but three years later those repairs were found to be inadequate. So the old structure was torn down and replaced with a new tower in 1999.[8]

Then, only two years later, the Maltese freighter *Kouras V* struck the new tower. The collision occurred shortly after midnight on January 22, 2001. The master of the vessel told the Coast Guard that high winds had forced his ship into the tower shortly after he had reduced speed to allow a pilot to board. The ship suffered two horizontal cuts in the hull—one above the waterline and one below. The tower was damaged significantly, but the damage was reparable.[9]

Six relatively peaceful years went by, and then inbound M/T *Axel Spirit*, registered in the Bahamas, struck the Ambrose Light at two o'clock in the morning on November 3, 2007. Initial reports said that high winds caused by Tropical Storm Noel might have been a key factor in the collision. According to the Coast Guard, the tower suffered "damage to its legs (and) stanchion, and the revolving light is bent and no longer rotating."[10]

It was beginning to become obvious that the Ambrose Light was no longer useful. It had become something of an impediment to navigation. There were better ways to mark the approach to the harbor, using sea buoys. The tower was a victim of both accidents and progress. The Coast Guard conferred with all the affected stakeholders including mariners, pilots, shipowners, tugboat companies, and barge operators. They reached a consensus: enough was enough. Whatever the cause of allisions—the wind or waves or mariner inattention—we cannot have ships strike the tower repeatedly. Thus, the Coast Guard reached the decision to dismantle the Ambrose Light after forty-one years of service. They installed a temporary light buoy some 300 yards in front of the damaged tower. Costello Dismantling Company of Middleboro, Massachusetts, was contracted to remove the tower, with assistance from a number of tugboats in July 2008.[11]

"It was determined that Ambrose Light is not absolutely necessary to perform safe navigation, through Ambrose Channel," said Chief Warrant Officer Darren Pauly of the Coast Guard. "Instead, several navigation aids will be added and Ambrose Channel will be extended to create a wider, more navigable channel for larger commercial vessels."

"Ambrose Light has had three allisions since 1996. That is the main reason it is being replaced. It is not cost effective to rebuild the Light every time it is struck. With our new improvements to the channel and newer Light-Emitting Diode (LED) technology, the optics on large buoys provide the same purpose and are easier to maintain," said Pauly.[12]

The new system has worked rather well right up to the present day, so let us follow the arrival of a typical large container ship as it makes it way from a foreign port to the APM Terminal in Port Elizabeth. Let us imagine that we are on board. We shall note significant landmarks on both sides as we ride along—including the lighthouses, the forts, and the anchorages.[13]

Meanwhile, there is a set of procedures that the captain must follow. Our ship must file a notice of arrival ninety-six hours in advance with the National Vessel Movement Center (NVMC) in West Virginia. This center, run by the Coast Guard, records and monitors the arrival of ships at all U.S. ports. The captain of the ship reports the name and home port of the ship, the estimated time of arrival, the cargo and crew onboard, as well as the recent ports of call.

This information is then forwarded to Sector New York, the largest Coast Guard operational field command on the East Coast, located at historic Fort Wadsworth on Staten Island, New York. At the Sector Command Center, the vetting duty officer reviews the information and determines which vessels need inspection. The officer then lets the Vessel Traffic Service know so that they do not let ships into the harbor if they need an inspection before entering. A very complicated grading system takes into account the national flag under which the ship is sailing, the vessel operator, the vessel owner, the vessel history, and the last port of call.[14]

Assuming that our ship is cleared to enterer the harbor, it will head for the Ambrose Sea Buoy, located some fifteen nautical miles southeast of Breezy Point, Queens. That buoy marks the entrance to New York Harbor. Nearby, the Sandy Hook pilots maintain a pilot vessel "on station" year round. Ships approaching the Ambrose Pilot Station from the sea are asked to give the pilot boat three hours' advance notice. Then they should make a second call to the pilot boat when they are one hour from the sea buoy. The pilot has detailed, specialized knowledge of the approach to the port. He will board the ship by climbing a ladder off the side of the ship. It becomes the pilot's job to ensure that the ship gets to its destination safely and on time. The approach begins by following the Ambrose Channel, the only shipping channel in and out of the harbor. Of course, the pilot must pay strict attention to the technical aspects of his job, noting all of the buoys and channel markers; however, we, as riders, have the opportunity to take note of the rich history on both sides of the channel as we ride along.[15]

As the Sandy Hook pilot comes aboard our ship, we can look way off to the left side and see the Navesink Twin Lights over on the New Jersey side, before proceeding up the Ambrose Channel. This is the first of a number of historic lighthouses we shall pass on our way into the terminal. To be sure, these lighthouses are no longer essential; rather they are of historical interest. The mariners of today rely on electronic aids to navigation. Lighthouse historians tell us that the Navesink Twin Lights facility is noted for being the site of many firsts. In the early days, this lighthouse was the first American landmark seen by incoming immigrants, even before they saw the Statue of Liberty. The location at Highlands, New Jersey, was ideal since it was the highest point on the Atlantic coast's mainland.

In addition, this lighthouse played a key role in the development of telegraphy. It was in 1899 that Guglielmo Marconi came to the United States to test his new wireless telegraph. He set up his equipment at the Navesink Twin Lights. It was used for the first time to report the victory of Commodore George Dewey in the Spanish-American War. Thereafter, it became the first station able to send and receive messages on a regular basis. Historically, it was also notable for being the first U.S. lighthouse to use a Fresnel lens. At one time, it had the most powerful beam of any American lighthouse; however, it was decommissioned in 1949.[16]

As we continue traveling from the Atlantic Ocean into Lower New York Harbor, we are passing the Sandy Hook peninsula that stretches some six and

FIGURE 7 The Sandy Hook Lighthouse in Sandy Hook, New Jersey; photo taken on February 6, 2019. It is the oldest working lighthouse in the United States, built in 1764 to aid mariners entering the southern end of New York Harbor. (Photo credit: John Van Decker / Alamy Stock Photo.)

a half miles over on our left side. Sandy Hook being located at the harbor's entrance made it a strategic defense site. As history-minded travelers, we know that on our port side we are passing Fort Hancock, the first of a number of silent sentinels along New York Harbor. The original U.S. Army granite fort, built between 1859 and 1867, was known simply as the Fort at Sandy Hook. In those days, there was always fear of an enemy invasion from the sea into coastal cities like New York. Later, in 1895, it was named Fort Hancock in honor of the Civil War hero, Major General Winfield Scott Hancock.

The fort was continually built up and expanded with coastal artillery in the twentieth century, right through both world wars. After 1945, however, it was evident that the real threat was from enemy airplanes and missiles, and coastal artillery was no longer needed for defense. However, a new use for the fort was found during the Cold War with the installation of antiaircraft guns; and this approach was quickly followed in 1954 by the introduction of Project Nike—a line-of-sight antiaircraft missile system that lasted until 1974, when the fort was decommissioned. Then, with the changing times, most of Sandy Hook became part of the Gateway National Recreation Area, administered by the National Park Service.[17]

As we continue our journey through history, just a bit north of Fort Hancock we can spot the Sandy Hook Lighthouse over on the left side—notable for being the nation's oldest operating lighthouse. It was built at the urging of forty-three New York City merchants, who had lost considerable money in 1761 due to shipwrecks. They raised the money from the proceeds of two lotteries. The first lottery in 1761 raised enough money to purchase four acres of land on Sandy Hook. The second lottery, completed in 1763, raised the money to fund the construction of the lighthouse. Built of rubble, the structure was octagonal, and it rose to a height of 103 feet. It was first lit on June 11, 1764. It guided ships around the treacherous shoals that had claimed so many vessels. The keeper resided in a small stone house next to the tower. His contract permitted the privilege of "keeping and pasturing two cows," but the contract also warned him against using the tower as a "public-house for selling strong liquors."[18]

Many years later, in 1857, a third-order Fresnel lens was installed, and it remains there today. In the early years, illumination was provided by whale oil and kerosene. Officials tried lighting the tower with electricity on an experimental basis in 1896, but it was not fully electrified until the 1930s. The tower was originally on the northern tip of Sandy Hook; but since that time, there has been a gradual buildup of land, so that the tower is now about one mile south of that point.[19] In 1964, there was a celebration to mark the two hundredth anniversary of the lighthouse. At that event, the Department of the Interior declared the lighthouse as a National Historic Landmark. Then, in 1996, the lighthouse was transferred from the Coast Guard to the National Park Service.[20]

As we continue traveling north in the Ambrose Channel, the next significant structure we find is the Romer Shoal Lighthouse, about two miles north of Sandy Hook on the left. It makes an interesting contrast with the Sandy Hook Lighthouse, which was purpose-built on dry land in 1764; Romer Shoal Light was erected in 1898, more than one hundred years later, and it was assembled, faster and cheaper, with entirely new methods. It was typical of the form known as a "sparkplug" lighthouse, based on its shape. It was built on a caisson base with prefabricated, standardized iron plates. This lighthouse was named after the pilot boat *William J. Romer* that sank here in 1863. It stands four stories tall, is painted red and white, and is equipped with a fourth-order Fresnel lens.[21]

Recognizing its significance, the Romer Shoal Lighthouse was added to both the New Jersey and National Registers of Historic Places in 2007. Later, a philanthropic executive from Staten Island purchased it for the benefit of

FIGURE 8 The West Bank Lighthouse; photo taken on September 15, 2019. Installed in 1900, it is located on the west side of the shipping channel in New York Lower Bay that leads to the Narrows and continuing onto New York Harbor. (Photo by Angus Kress Gillespie.)

the National Lighthouse Museum, which was to use it for tours and overnight visits. The plan made sense because, at that time, the lighthouse was in good condition. Tragically, the lighthouse was severely damaged by Superstorm Sandy in 2012, with the doors and windows on the first two levels blown out and the interiors completely flooded. The riprap foundation was reduced; and, as a result, it is now difficult to access the property. As this is written, a nonprofit foundation is trying to raise funds to repair and restore this historic structure.[22]

As we proceed up the Lower Bay, the next lighthouse that we encounter on the port side is West Bank Lighthouse. It is of similar sparkplug design that we have just seen, but it is considerably larger, with a height of 69 feet, as opposed to only 54 feet for Romer Shoal.[23] Work began on the construction of the West Bank Light in July 1900. By this time, there was a standard way of doing things that saved time and money. Workers would first build a large cylindrical iron base, or caisson, at a foundry in upstate New York, and tow it to the site, which would have been already excavated. They would then sink the caisson and proceed to fill it with concrete. The sunken base would be surrounded by some 250 tons of riprap, or loose stone used to protect the structure. The prefabricated superstructure could then be placed on the base, and topped off with the lantern deck. The whole process took only six months—it was completed in December 1900. The fixed white light of the fourth order was lit on January 1, 1901.[24]

For eight decades, the West Bank Lighthouse was continuously manned, but the Coast Guard automated it in the early 1980s. By May 2007, the Coast Guard decided that the light was no longer needed, and it was declared excess. As a result, it was offered at no cost to eligible entities, but there were no takers. Thus, the General Services Administration (GSA) conducted an auction during the summer of 2008. There was a winning bid of $245,000, but the sale was never finalized. On June 30, 2010, the GSA again put the lighthouse on the auction block. This auction closed on September 28, 2010, with a winning bid of $195,000. The winner turned out to be Sheridan Reilly, who purchased it out of an altruistic sense of duty. As a boater, he had passed it many times, and he felt obliged to do his part to save this historic structure.[25]

Armed with a pair of binoculars, as we continue heading north, looking way over to the port side, there is a sad sight—the ruins of the Old Orchard Shoal Lighthouse. The deadly Hurricane Sandy destroyed it on October 29, 2012. The lighthouse was swept away by a storm surge, in this case a 32.5-foot wave, as reported by the National Oceanic and Atmospheric Administration (NOAA). All that is left is the concrete platform and a pile of riprap.[26] Originally completed and lighted in 1893, it served its purpose for a full 119 years. It was built in the familiar sparkplug design, mounted on an iron caisson filled with concrete. It was placed in an area of Lower New York Harbor that used to be called Princess Bay. It stood in 17 feet of water, and it had a fourth-order lens 51 feet above the water.[27]

At the same time that the West Bank Light was excessed by the Coast Guard in 2007, so was the Old Orchard Shoal Light. Again, it was offered at no cost to eligible entities. In the same way that no qualified group was found, the GSA proceeded with the process of auctioning off the lighthouse, beginning in the summer of 2008. Because of many legal complexities, the first two auctions were not finalized. Finally, the third auction ended on December 1, 2010, with the winning bid of $95,000. Of course, as it turned out, the winner was able to enjoy ownership for a period of less than two years. Shortly after the Hurricane Sandy disaster, on December 4, 2012, at a fundraising event, Linda Dianto, executive director of the National Lighthouse Museum, made an announcement. She revealed ambitious plans to recover and reconstruct the lighthouse on the museum grounds, the former site of the U.S. Lighthouse Service's General Depot in St. George, Staten Island.[28]

Some seven years later, I asked Dianto about those plans. She explained fragments of the lighthouse had been scattered by the storm surge. Initial investigators found some fifty-three possible points of interest on the ocean floor. Fundraising for the project was difficult. With the passage of time, some of the fragments were covered over with sand. At the present time, there are no plans to reconstruct the whole structure from the scattered fragments. However, it is entirely possible that the museum can retrieve some of those artifacts and put them on display. There is also a $35 million planned expansion of the museum, which may include a reconstructed replica of the original Old Orchard Shoal Lighthouse.[29]

Moving along, we know that we are passing from Lower New York Bay to Upper New York Bay, when we see Coney Island, Brooklyn, off to the starboard side. This celebrated playground is well known for its crowded beaches and its boardwalk with cotton candy, funnel cakes, and roller coasters. From the channel, the most obvious landmark, especially at night, is the illuminated Parachute Jump. It was originally built for the 1939 World's Fair in Queens, but it was moved to Coney Island in 1941. Unfortunately, the ride stopped operating in 1968. However, the structure was renovated and was lit with some 450 lights in 2006. It was a good start, but Brooklyn borough president Marty Markowitz was unhappy with the lighting, which was minimal. He proceeded to spearhead a movement to improve the lighting with a $2 million upgrade and 8,000 LED lights. On June 21, 2003, Markowitz and amusement park officials threw the switch on the new lighting—so bright that it can be seen from space.[30]

In contrast to the brightly lit Parachute Jump is the relatively subdued Coney Island Lighthouse, located at Norton's Point on the west end of Coney Island. It was built in 1890 with a central, narrow steel column with 87 steps. The tower stands 61 feet tall with a lantern room on the top. It was automated in 1989 and remains operational. Unlike our classic view of an isolated lighthouse situated on a remote lonely peak, this lighthouse is surrounded by a highly built-up neighborhood. The lighthouse has an associated keeper's dwelling, with two floors and an attic.[31]

The last resident keeper was Frank Schubert (1915–2003). In addition to performing his regular maintenance duties, he went out of his way to save the lives of fifteen sailors. For his extraordinary service, President George H. W. Bush recognized him at the White House in 1989. He continued working until December 11, 2003, when he died at the age of eighty-eight, the last of America's lighthouse keepers. His career had lasted sixty-five years, with forty-three of them at the Coney Island Lighthouse.

"The Coast Guard mourns the loss of its most courageous sentry of the sea, said Captain Craig T. Bone, commander of Coast Guard Activities New York, after Schubert died. "His devotion to duty and courage are unequaled."[32]

Somewhat farther along, on the port side, there are two rather bleak manmade islands that most of us have never heard of. These two islands are west of our main ship channel and about a mile off the Staten Island shore. The construction of the two islands was authorized by the New York State legislature in 1866 to monitor arriving immigrants for the purpose of preventing the bringing in and spreading of infectious diseases by providing places of quarantine. Vessels that had passengers who were suspected of having smallpox, yellow fever, or other contagious diseases were ordered to stop on their way to the city. Building the two islands cost about $400,000 and took a full six years. Rocks were barged down the Hudson River and dumped. The islands were further built up by pumping in large amounts of sand.[33]

The larger of the two islands is Hoffman Island, named for John T. Hoffman, who was mayor of New York City from 1866 to 1868 and governor of New York State from 1869 to 1871.[34] Hoffman Island is about nine acres in size, and it was used as a "Quarantine of Observation" for people who had been exposed to contagious diseases. The smaller of the two islands is Swinburne Island, named for John Swinburne, who oversaw its construction.[35]

It is about three acres in size, and it was used for the actual cases. The two facilities were in use for some forty-nine years, from 1866 to 1915, during which thousands were detained and hundreds died. As the number of immigrants fell in the twentieth century, there was no longer a need for these two quarantine islands, so in 1921 the federal government purchased them from the State of New York.[36]

Beginning in 1938, a new use was found for Hoffman Island. It was taken over by the U.S. Maritime Commission for the training of new seamen in the merchant marine. It was a place with few distractions, ideal for an intense curriculum. There was an urgent need for them to serve in a greatly expanded service. They were needed to help transport large quantities of war materiel, supplies, and troops across the oceans of the world. At this island base, seventeen buildings housed the six hundred men and sixty Coast Guard officers and enlisted men who ran the school. The trainees signed up for a three-month course that covered seamanship, engineering, and navigation. Every week fifty new trainees were admitted and fifty new graduates were sent to sea. The Maritime Commission was responsible for the maintenance of the infrastructure of the island; however, the Coast Guard was responsible for the training of the men. Although the men were being trained for *civilian* jobs as merchant seamen, the training was paramilitary and rather strict.[37]

By 1947, with the end of World War II, the maritime school was no longer needed, and it was officially closed. No new use was found for Hoffman Island, and it has been off-limits to the public since 1947. The two islands are now managed by National Park Service as part of the Gateway National Recreation Area for the benefit of seabirds, including the snowy egret, the black-crowned night heron, and the great black-backed gull.[38]

Coming up next on the starboard, or Brooklyn, side is the Gravesend Bay Anchorage. Geographers tell us that Gravesend Bay is an estuary—a partially enclosed body of water, where freshwater reaches ocean seawater.[39] To borrow an analogy from automobile driving, it is a good place to pull over and park. Of course a ship does not park but drops its anchor. To take advantage of this convenient anchorage, there are four commonsense federal regulations. First, the ship must maintain a bridge watch. Second, the ship must guard and answer Channel 16, the distress channel, and must maintain an accurate position plot. Third, the ship must notify the Coast Guard on Channel 12 for Port Operations, when beginning and ending the transferring of petroleum products. Fourth, the ship must notify the Coast Guard on Channel 12 when anchored and when leaving the anchorage.[40]

There are at least four good reasons why a ship might choose to anchor here. First, a ship might need to refuel, or it might need a fresh supply of stores. Second, a ship might need to offload cargo. Of course, the ideal place to offload cargo would be a regular terminal at a regular port; however, there are circumstances when that is difficult, such as if the water level is too low at the terminal for the ship to approach. In such a case, a tanker might offload oil onto a barge, pulled by a tug, at the anchorage, or a bulk carrier might offload grain, coal, or gravel onto a barge. Third, a ship might need some kind of emergency repair before it can proceed, such as restoring a loss of power or fixing the control mechanism of a rudder. Fourth, a ship might choose to anchor rather than continue when the captain encounters high wind or dense fog. In such a case, he might decide that it is too dangerous to proceed.[41] As we pass Gravesend Bay on the starboard side on our way to Port Elizabeth, looming ahead is the massive Verrazzano-Narrows Bridge. As we have been moving along, I have pointed out a few significant landmarks. To see these, for the most part, we have to know where to look. However, for the Verrazzano-Narrows Bridge, it is obvious even for the most inattentive observer. The

FIGURE 9 View of Fort Wadsworth, Staten Island, New York, and the Verrazzano-Narrows Bridge, which connects Staten Island to Brooklyn, New York; photo taken on May 26, 2019. (Photo credit: Robert Quinlan / Stockimo / Alamy Stock Photo.)

crowning achievement of builder Robert Moses, it was built to connect Brooklyn on our starboard side with Staten Island on our port side.[42] Construction of this engineering marvel began in April 1959, and it was opened to traffic on November 1, 1964. At that time, it was the world's largest suspension bridge, spanning 4,260 feet between two 690-foot-high towers. Today, there are 143,000 miles of cable bearing the weight of twelve traffic lanes—six on each level. Construction required more than 600,000 cubic yards of concrete. The completed project stretches for some 13,700 feet, or about the length of fourteen football fields.[43]

For the mariner, the Verrazzano-Narrows Bridge is more than a bridge; it is an important marker. We have now traveled 13.6 miles from the Ambrose Sea Buoy. This area is known as the Narrows, a body of water less than a mile wide. We have now passed from Lower New York Bay to Upper New York Bay. Shortly after we pass underneath the bridge, we reach the point where the Sandy Hook pilot is relieved by the harbor pilot. The job of the Sandy Hook pilot has been to bring the ship from the Atlantic Ocean to the harbor. It will now be the job of the harbor pilot to maneuver the ship through narrow waterways that lead to the terminal and to supervise the docking of the vessel with the help of tugboats. With the harbor pilot at the helm, we now have 9.3 miles to go to get to the dock at Port Elizabeth.[44]

Coming up quickly on the port side is Fort Wadsworth on Staten Island. It was a natural site for the mounting of guns to be used against hostile ships.[45] The fort has a long history, starting with the Dutch, who fortified the site in 1636. The British took it over from the Dutch in 1664. After the American Revolution, the U.S. government took over and expanded it from time to time, building Fort Tompkins in 1807 and Fort Richmond in 1847. In time, the complex was named Fort Wadsworth in honor of Brigadier General James S. Wadsworth, who was killed in action in the Civil War at the Battle of the Wilderness.

Throughout the first half of the twentieth century, the coastal artillery batteries were maintained and augmented at Fort Wadsworth. However, after World War II, with the advent of jet aircraft and guided missile systems, the mission of the fort began to change. A Nike missile antiaircraft system was installed in 1952, and it continued there until 1964. There was also a four-gun, 120-millimeter antiaircraft battery located at the fort between 1952 and 1955. The fort became the headquarters of the New York Naval Station from 1990 to 1994.[46] Then, in 1994, the U.S. Navy turned Fort Wadsworth over to the National Park Service.[47]

Coming up almost immediately on the starboard side is the sister fort to Fort Wadsworth, Fort Hamilton, surrounded by the community of Bay Ridge in Brooklyn. The U.S. Army began construction of the one-acre fort in 1825, and it was completed six years later. Built of gray granite, it was designed to defend against attacks from both water and land. During the fort's long history, many army officers rotated in and out of this duty station. At least three of them achieved subsequent fame. First, in 1841, Robert E. Lee, best known to history as the commander of the Confederate Army, was assigned to Fort Hamilton as post engineer. In this five-year tour of duty, he improved this and other nearby military installations. Second, we note that Thomas "Stonewall" Jackson, remembered as one of the most effective Confederate generals, came to Fort Hamilton after graduating from West Point in 1846. After distinguishing himself in the Mexican-American War (1846–1848), he returned to Fort Hamilton with the rank of major. Third, Major General Abner Doubleday, known for developing the game of baseball, served as post commander at Fort Hamilton at the beginning of the Civil War in 1861.

Over time, the fort grew in size. In 1891, it was enlarged to 96 acres. In the twentieth century, it grew to 155 acres. As the fort grew, its mission changed from coastal defense to personnel processing. In both world wars, Fort

Hamilton became a center for embarkation—processing the movement of troops from the United States to overseas commands, all the mean while making sure that they were properly equipped. The fort also served as a center for army separation where soldiers were officially released from active duty after completing their term of service. If all went well, they would receive an honorable discharge.[48]

Fort Hamilton has evolved into a joint base serving the needs of active-duty Army, Navy, Air Force, Marines, and Coast Guard personnel, and their reserve components. Much of what happens here is what we in the civilian sector would call human resources management and benefits administration. Of course, the military has its own way of doing things, so these functions have somewhat different names. A quick look at the Fort Hamilton website gives an idea of these services. It includes such things for new people as "Recruiting Battalion" and "Military Entrance Processing Station." For active-duty personnel we find categories such as "Beneficiary Counseling Assistance" and "Morale, Welfare, and Recreation."[49]

Shortly after passing Fort Hamilton on our starboard side, we come upon the Stapleton Anchorage over on our port side. This is a busy and desirable piece of waterway real estate because of the limited amount of deep-water anchorage space in the Upper Bay of New York Harbor. Years ago, when there was less maritime traffic, ships could pull over to the anchorage and stay as long as they wished. However, in recent years, the number of ships has increased and these ships have become bigger. As a result, mariners often found the anchorage filled to capacity.

There was a growing consensus in the maritime industry that something had to be done because having enough anchorage space is necessary for the overall safety and economic success of the port. Often, because the available anchorage space was already taken, many ships were forced to anchor outside the port at the Ambrose meeting point—often for days, while waiting for their turn at anchorage space. These vessels were unable to finish their necessary business—for such things as taking on a new supply of food or offloading cargo onto barges or taking on more fuel. Even worse, the lack of anchorage space created a safety risk for any ships that might need to anchor in an emergency, such as loss of power or steering.[50] As a result, in 1976, the U.S. Coast Guard imposed a forty-eight-hour limit for anchoring in the Stapleton Anchorage.[51]

Recently, a further improvement was made to improve the efficiency of the Stapleton Anchorage. The Sandy Hook pilots proposed a scheme to create assigned berths within the anchorage—comparable to the painted lines in a

FIGURE 10 Staten Island Ferries docked at the St. George Ferry Terminal on Staten Island. The system provides a link between Staten Island and Whitehall Ferry Terminal in Lower Manhattan. (Photo taken on September 15, 2019, by Angus Kress Gillespie.)

parking lot. Of course, lines cannot be painted on the water. Instead, officials created an electronic overlay for the navigational charts to be published by the National Oceanic and Atmospheric Administration (NOAA) Office of Coast Survey. The overlay is reminiscent of the familiar bingo card, consisting of parallel vertical and horizontal lines that are labeled and charted over the anchorage area. The scheme created a large number of "parking spaces" or "anchor boxes," each 600 feet long and 300 feet wide. The plan made so much sense that it was readily adopted by the U.S. Coast Guard Sector New York's Vessel Traffic Service (VTS).

"Incorporating these overlays in an ENC [electronic navigation chart] will increase safety and efficiency in the port's limited anchorage space. VTS will be able to direct a vessel to a specific grid location, and that vessel will be able to see the location on their electronic chart system," said USCG Captain Michael H. Day, captain of the port, Sector New York.[52]

Coming up quickly on the port side, we can see the big orange boats of the Staten Island Ferry docked at the St. George Ferry Terminal on Staten Island. The system provides a link between Staten Island and the Whitehall Ferry Terminal in Lower Manhattan. This ferry reminds us of the early days, before bridges and tunnels, when ferries were the only connection between

Manhattan and the outer boroughs. Today, the Staten Island Ferry serves 22 million people per year, at the rate of 70,000 a day on weekdays, fewer on weekends.[53]

On the one hand, many of the passengers are residents of Staten Island who use the ferry to commute to work in Manhattan. Some say that they have the best of both worlds—relatively affordable housing in Staten Island and good-paying jobs in Manhattan. On the other hand, some of the passengers are simply tourists. All of the New York City guidebooks say nice things about the Staten Island Ferry. While bored commuters are reading the newspaper or checking their cellphones, eager tourists are lining the rails to see the Statue of Liberty, taking full advantage of the five-mile and twenty-five-minute ride to enjoy the sights of the busy harbor. Since the ferry runs twenty-four hours a day and seven days a week, some of the tourists schedule a romantic ride at night when they can admire the bright lights of the big city.[54]

Shortly after passing the St. George Ferry Terminal, our ship will make a sharp left turn into the Kill van Kull, as named by the original Dutch settlers. It is a tidal strait with Staten Island, New York, on our left or south side, and Bayonne, New Jersey, on our right or north side. The Kill van Kull is a three-mile-long passageway that will take us from the Upper New York Bay to our destination in Newark Bay. Because this waterway is both very narrow, only about 1,000 feet wide, and very busy with both tankers and container ships, our harbor pilot will have to be especially attentive.[55]

Our trip through the Kill van Kull presents us with a heavily industrialized landscape on both sides. As we enter from the eastern end, we quickly come upon the Atlantic Salt Company on our port, or Staten Island, side. Every fall, this waterfront depot gets huge shipments of salt offloaded from bulk carriers. The salt is leveled by a bulldozer and then covered with a huge tarpaulin. The New York City Department of Transportation uses the salt to treat the roadways following snowstorms. The company has won high praise for allowing the Staten Island community to use the facility in the off season for art festivals, when the buildings there are lit with video and light installations.[56]

Next, coming up on our starboard side, we find the sprawling facilities of IMTT (International-Matex Tank Terminals) in Bayonne. This company handles bulk quantities of petroleum products, commodity chemicals, vegetable and tropical oils, oleochemicals, biofuels, and ethanol. It is a huge facility hosting 620 tanks with 16 million barrels of total capacity, ranging in size from 500 to 250,000 barrels. The company enjoys highway, rail, and pipeline access as well as having seven recently dredged deep-water tanker berths and eleven barge berths.[57]

For IMTT in Bayonne, there are some humorous plays on words used by harbor pilots. On the one hand, they might say that IMTT stands for "I am too tired." Of course, that's just a joke, since the harbor pilot has to do his job tired or not. On the other hand, they might say that IMTT stands for "I missed the tide" or "I missed two tides." These sayings are no longer merely jokes; rather, they are expressions of occupational folklore, passed along verbally from one generation of pilots to another.

To understand what is being said requires a bit of explanation. Many of the docks at IMTT are what are called "finger piers." These piers are perpendicular to the shore; in other words, they stick out into the water like the fingers on your hand. These piers stand in distinction to parallel piers, which are simply alongside the shore. The finger piers present a challenge to docking pilots, especially during either high tides or low tides, when the water is flowing either in or out. The flow of the water stresses the situation when the pilot is trying to insert the tanker into the finger pier. What the pilot wants to do is to hit the sweet spot called "slack water," that brief period of time when there is no movement of water either way in the tidal stream. So now we can understand the real disappointment behind the saying "I missed the tide." It means that a relatively easy job has become very difficult. As one harbor pilot told me, "You can't fight Mother Nature."[58]

Meanwhile, on the port side, we see the Greek Revival buildings of Sailor's Snug Harbor. This was a retirement home for "aged, decrepit, and worn out seamen." It was created by Robert Richard Randall, who had made a fortune in the maritime industry. In his will, written in 1801, he left his property to create this facility on Staten Island. It was opened in 1831 with twenty-seven residents, and it was one of the first retirement homes in the United States. By 1900 it had one thousand residents. After World War II, there was a decline in enrollments, and in time the facility was moved to Sea Level, North Carolina. Today, it is a cultural center for the residents of Staten Island and the rest of New York.[59]

Next on the port side, we see Caddell Dry Dock and Repair Company. They offer floating dry dock services, an interesting process that involves a floating chamber having walls and a floor that can be flooded to allow a vessel to be floated in. Then all the walls are closed, and the water is pumped out, allowing the vessel to rest on a dry platform. Once this is accomplished, hull maintenance, including scraping, sanding, pressure washing, and painting, can proceed. In addition, machinists can carry out repair work such as changing propellers and repairing rudders. Besides dry dock services, Caddell does pier-side repairs as well. It is a busy facility with eight piers, each equipped with cranes able to extend 200 feet and capable of handling loads of up to 6,500 tons.[60]

A few years ago, Wendell Jamieson, a reporter for the *New York Times*, wrote of the Caddell yard, "You can hear the sounds of ships being scraped and repainted with spray guns and banged back into shape, and the shouts of men who do this. Forklifts beep as they back up; crane motors whir. Twice a day, you can hear a whistle blow at Caddell's—once for lunch, once for quitting time."[61]

Moving along, we see on the port side the Moran Tugboat Yard. It is easy to recognize because there is almost always a cluster of tugboats at the dock. In addition, we should mention that there is a custom in the tugboat world of each company having its own livery, or a special pattern, or paint scheme on the exterior of its boats. This makes the tugboats readily identifiable. In the case of Moran, the pattern is to paint the deckhouse, or superstructure, with a distinctive maroon paint and to paint the hull with a distinctive chrome green.[62] At the same time, the black smokestacks feature a large white letter "M." Historians tell us that it was Michael Moran, the founder of the company, who himself painted that first white "M" on a tugboat smokestack around 1880.[63]

As of this writing, Moran has a fleet of twenty-nine tugboats assigned to New York Harbor. These fall into three broad categories—tractor tugboats, conventional twin screw tugboats, and articulated tug-barge combinations (ATBs). As we are passing by the Moran Yard, an astute observer can point out the different types.[64]

Only four of them are the so-called "tractor tugs." These tugs are relatively new, very powerful, and equipped with the latest technology. They are especially suitable for maneuvering and docking very large vessels. What makes such a tugboat special is that it is equipped with what is popularly known as a "Z-drive" or technically as an "azimuth thruster." This is a type of marine propulsion unit that can rotate 360 degrees, making for rapid changes in thrust direction. There is no need for a conventional rudder. These tugboats are very useful and very expensive. A new one can cost upward of $12 million.[65]

In addition, Moran has fifteen conventional twin screw tugboats, equipped with diesel engines and two propellers. These are the familiar workhorses of the harbor. They can tow a ship or a barge, and they can help in the maneuvering of a ship to or from a berth. They can be called upon for emergency towing if a vessel loses propulsion power. They are also useful for escort towing—a precautionary measure to accompany a vessel while

FIGURE 11 Tugboat passing under the Bayonne Bridge on the Kill van Kull. The Bayonne Bridge is the fifth longest steel arch bridge in the world, and it connects Bayonne, New Jersey, with Staten Island, New York. Photo taken on October 8, 2007. (Photo credit: Ellen McKnight / Alamy Stock Photo.)

navigating in the harbor to protect against damage or collision in the event of an engine or steering failure.[66]

Finally, Moran has ten articulated tug barges (ATBs). This is a specialized tugboat that always comes with its own barge. The barge, of course, by definition is a marine vessel used to carry cargo, but it is not independent since it has no propulsion. It must be pushed along by its tug, easily recognizable because it has a second, elevated wheelhouse to allow the operator see over the barge in front. Years ago there was a predecessor, the integrated tug barge (ITB). This combination featured a rigid connection between the tug and the barge, creating basically a single unit. With the ATB combination, there is a hinged connection, which allows for some separate movement between the two parts. The barges for these ATBs may carry diesel fuel, gasoline, urea, or grain. The ATB combination can be quite large, which calls for careful navigation. For example, you might have a barge that is 500 feet long, pushed by a tug that is 120 feet long.[67]

As we pass by the Moran yard, we are now less than a mile away from the Bayonne Bridge that looms ahead. When it opened in 1931, this bridge connecting Staten Island and Bayonne was considered to be an engineering marvel. The roadway of the bridge was a full 151 feet above the Kill van Kull, allowing passage of the biggest ships of the day. The Bayonne Bridge worked just fine for seventy-five years; but then in 2006 there was an alarming development—the citizens of Panama voted to expand the Panama Canal. The new canal would be able to accommodate much larger ships. The existing canal serviced container ships with about 5,000 standard-sized shipping containers, while the proposed new canal could carry container ships carrying some 14,000 containers. All of a sudden, the Bayonne Bridge was obsolete, and something had to be done.[68]

The situation was perhaps best summed up by engineer Beth DeAngelo: "The expansion of the Panama Canal has been a game changer for global goods movement. Without changes to the Bayonne Bridge, an increase in Panamax shipping would have diminished access to our goods and our community. We stood to lose a significant amount of shipping business to other East Coast ports."[69]

So the Port Authority of New York and New Jersey undertook a $1.3 billion project to raise the roadway of the bridge by 64 feet, to a height of 215 feet above the water. At first, engineers considered demolishing the existing bridge and then building a new and better bridge, but that idea was quickly dismissed. Instead, they decided on something unprecedented and very ambitious—they would raise the existing bridge roadbed without shutting down traffic and without disrupting the shipping channel. Engineers began designing the project in 2011, and construction began in 2013.[70]

How was it possible to raise the height of the bridge without closing it to traffic? It had to be done one step at a time. First, the traffic was shifted to the western side of the roadway, allowing one lane going in each direction. Then, the roadway on the eastern side was removed, and the concrete piers there were replaced with taller ones. Next, giant cranes could lift and put into place a higher roadway on the eastern side. Then, the process could be repeated for the western side.[71]

The Orient Overseas Container Line (OOCL) is a Hong Kong–based container shipping line. It was in July 2017 that their ship, the OOCL *Berlin*, passed under the recently raised Bayonne Bridge. It was a significant moment. This was a vessel with a capacity of 13,200 standard containers, the largest ever to enter the Port of New York and New Jersey. The ship was headed for the Maher Terminal in Port Elizabeth, having begun its trip in Hong Kong and having passed through the Panama Canal.

Captain Andrew E. Glassing, a Sandy Hook pilot who helped guide the OOCL *Berlin* into the port, said, "It was first and foremost, an extremely humbling experience. I don't think there was anything that we hadn't

FIGURE 12 *Bruce A. McAllister* tugboat passing under the Bayonne Bridge on the Kill van Kull, showing the roadway of the bridge having been raised by 64 feet, to a height of 215 above the water. The tug is returning to the McAllister yard after completing a job. Photo taken on July 9, 2020. (Photo credit: Rev. David Rider.)

anticipated. It was kind of neat because as we went under the new Bayonne Bridge you could see all of the workers, they were standing there, kind of saluting the ship," he said, referring to the workers who were finishing up the removal of the lower level of the bridge.[72]

As we pass underneath the Bayonne Bridge on our imaginary voyage, we soon come upon Bergen Point on the north side of the Kill van Kull, and it marks the spot where we take a hard right turn into Newark Bay. At this point we have traveled 21 miles from the Ambrose Sea Buoy. We now have less than a mile to go to reach our destination, the APM terminal in Port Elizabeth.

6

Navigation

●●●●●●●●●●●●●●

Ships entering or leaving Port Newark or Port Elizabeth must pass through New York Harbor, which is at the mouth of the Hudson River where it empties into New York Bay. It is not an easy task for a navigator because of the high traffic and strong tidal currents. The navigator must be ever watchful. The large oceangoing cargo ships have to be extra vigilant because the harbor is clogged with tugboats pushing barges. In addition, visibility is often limited because of fog, especially in the late spring and early summer.

Just as an automobile driver must know the rules of the road, so ship handlers must know seagoing traffic rules. They must know who has the right of way. They must know all about navigation lights and sound signals. Of course, in recent years ship handlers, like automobile drivers, have been greatly helped by the use of GPS. However, veteran navigators know better than to simply rely on their GPS, especially in the tight quarters of a harbor approach. They also rely on "aids to navigation" (ATON), including buoys of many shapes and colors used to mark the sides of channels. ATON also include fixed aids, or daymarks, which are often on pilings over the water, sometimes placed on shore.

Maintaining these buoys and daymarks is a task that falls to the U.S. Coast Guard, which functions in New York Harbor much like a state highway commission on land. Normally, when we think about the Coast Guard, we think of its highly visible missions. The first thing that comes to mind is the search and rescue mission. Rendering aid to people in distress and saving lives is a key mission of the Coast Guard. We often see heart-warming stories of saving lives at sea on the evening television news. Similarly newsworthy are the missions of drug and migrant interdiction. By way of contrast, the ATON mission is easily overlooked. It is not glamorous, and it never makes the evening news.

The ATON mission is dirty, dangerous, repetitive, and unappreciated. Yet, strangely enough, the Coasties I have talked with take a great deal of pride in their work. Why is this? Let us briefly follow a typical daily task of a buoy tender vessel. Buoys have to be routinely and regularly replaced. The old buoy is removed from its anchor chain, and the new buoy is attached to the chain. This procedure is necessary because, over time, the paint fades and the buoy is overcome by the accretion of marine growth, in addition to the normal wear and tear of corrosion. Simply put, the buoy is no longer highly visible. In a typical evolution, the ship comes alongside the buoy in need of replacement. The crew hooks it at the top, and they attach a cable. Then, they use a big crane to bring the buoy aboard and knock off some of the accretions. Finally, they use the same crane to put out a replacement buoy. The satisfaction comes from seeing the old, tired buoy replaced with a newly refurbished, shiny one.

Another reason that the Coasties who work in ATON take such pride in their work is that it so very dangerous. Swinging a buoy that may weigh up to 20,000 pounds onboard the ship without injuring or killing a crew member is a tricky proposition. It requires both skill and teamwork. Some have even called it the "Most Dangerous Job in the Military."[1] Granted, that claim may be somewhat exaggerated. Certainly, there are those who would argue the point. How about an aviation survival technician in the Coast Guard, known as a rescue swimmer? They jump out of helicopters to save those in danger of drowning in icy waters. How about an infantry rifleman in the Army? Surely, that is a dangerous job, especially in wartime. Let us grant that working on a buoy tender may not be the "most dangerous job in the military." Let us just say that it is one of the most dangerous.

In trying to understand the high morale of ATON crews, I consulted an unofficial online discussion group called "United States Coast Guard Forum." There I found an interesting exchange. Someone wrote, "I'm curious to know what it is like to work on a tender?" A crew member quickly responded: "Absolutely awesome in my opinion. We are 100' long, so it's a real small crew. Tight family here. We go out, work our tails off on the buoy deck and out in our small boat, we throw chain and heavy stuff around all day long, climb up and down ladders on shore aid towers, work super long days, and then play real hard. It's super awesome blue collar hard work, but with that comes a sense of real accomplishment and tight morale among crew."[2]

FIGURE 13 USCGC *Katherine Walker* (WLM-552), a 175-foot buoy tender homeported in Bayonne, New Jersey, transits beneath the Verrazzano-Narrows Bridge on March 8, 2016. The vessel's primary mission is maintaining over 300 floating aids to navigation in and around New York harbor and its approaches. (Photo credit: Petty Officer Third Class Frank Iannazzo-Simmons, USCG.)

As this is written, the buoy tender working the New York–New Jersey Harbor is the U.S. Coast Guard cutter *Katherine Walker*. It is classified as a coastal buoy tender. As such, it is relatively small, at 175 feet in length, compared with seagoing buoy tenders, which run up to 224 feet in length. The ship operates safely with a small, tight-knit crew of two officers and twenty-two enlisted personnel. The ship is not designed for speed. At 80 percent of power, it can travel at only about 12 knots. Instead, the ship is designed for maneuverability. It is equipped with the so-called Z-drive, a type of marine propulsion unit that rotates 360 degrees. This system allows for rapid changes in thrust direction. Thus, for example, the ship can pull alongside a buoy and maintain a steady position without anchoring.[3]

How did this ship get the name "Katherine Walker"? To find the answer, we have to go back to the 1990s when the Coast Guard built some fourteen new coastal buoy tenders. These ships were needed to replace the older buoy tenders dating back to the World War II era. Whenever you have a group of ships with a similar design, they make up a so-called class of ships. A decision was made to name this class of ships after historic, if not famous, keepers of lighthouses. Thus, all of these ships belong to the "keeper class." It was a kind and generous decision because, in the modern era, lighthouses have become automated and no longer require keepers. Naming these ships after lighthouse keepers is a way of giving young Coast Guard sailors a sense of history and pride in their service.

The story of Katherine Walker is a remarkable one. Born in Germany in 1846, she came to America and settled in Sandy Hook, New Jersey. There she met John Walker, who gave her free English lessons. They fell in love and got married. Later, John was appointed keeper at Robins Reef, located on the west side of Upper New York Bay, near the entrance to Kill van Kull. For a time, they had a happy life together, keeping that light working. Then, everything changed after her husband's death from pneumonia. Katherine replaced her husband, and she served as head keeper from 1894 to 1919, retiring at age seventy-three. Although she stood less than five feet tall and weighed less than one hundred pounds, she was given credit for saving at least fifty people from drowning.[4]

I wanted to learn more about the workings of this buoy tender, so I reached out to Lieutenant Torrey Jacobsen, the commanding officer of the vessel. He kindly invited me aboard during a regular weekday, after the ship had completed its regular duties and was tied up at the dock in Hoboken, New Jersey. Jacobsen showed me around the ship. I felt privileged to see the inner workings of a ship that few others will ever see. After a brief tour, we retired up a ladder to the captain's cabin, comfortably seated, as I conducted an interview with the recorder running. Torrey cheerfully answered my questions, and he began our long interview by explaining what he saw as the importance of his mission.

In his view, and I am inclined to agree, the *Katherine Walker* is of vital importance to the ports of New Jersey and New York. There is currently over $200 billion worth of commerce coming into these ports every year. If the channel were to be blocked, or if commerce were somehow otherwise disrupted, the adverse economic impact would be massive. This little ship with a crew of only twenty-four is responsible for lining up buoys marking the channel coming into the harbor. He said, "You might think that, without those buoys, driving a ship into the harbor would be just fine, but that is not true at all." He went on to explain that there are many shallow spots coming into the harbor. If there were some disruption to GPS, those buoys would ensure that the vessels could get in safely. With those buoys, commerce would continue to flow, with or without GPS or a navigation computer or a Sandy Hook pilot on board or Vessel Traffic Service (VTS) operating. Thus, the *Katherine Walker* plays a vital role.[5]

FIGURE 14 Seaman Chance Johnson, USCG, a buoy deck crewmember aboard the USCGC *Katherine Walker*, scrapes barnacles and marine growth from a buoy while under way in New York Harbor on March 8, 2016. (Photo credit: Petty Officer Third Class Frank Iannazzo-Simmons, USCG.)

I was curious about how Torrey got started in this line of work. He was happy to explain: "I actually grew up in the Annapolis, Maryland, area and so the Naval Academy was essentially in my backyard. I always wanted to go to the Naval Academy. I saw the midshipmen walking around Annapolis in their nice white uniforms, and I thought that was so cool." He went on to explain that he was in the Boy Scouts at the time. During our interview, Torrey was too modest to tell me that he had made the rank of Eagle Scout. I learned that only later. I now see that accomplishment as significant because it involved a service project that the scout plans, organizes, leads, and manages. In retrospect, it showed an early tendency for leadership. Interestingly enough, it turns out that his scoutmaster was a helicopter pilot for the Coast Guard. The scoutmaster encouraged him to consider serving in the Coast Guard. About the same time, he met a few Coast Guard enlisted people, and he was struck by their dedication, by their commitment to helping people. Then, he took a tour of the Coast Guard Academy and never looked back.

I asked Torrey about his experience at the Coast Guard Academy in New London, Connecticut. He explained that he took hours and hours of classes, much like any other college student. But there was a big difference. When the school year ended, most normal college students would go home. By way of contrast, Coast Guard cadets are sent to Coast Guard units for a twelve-week period. He explained that six weeks might be on a ship, and another six weeks might be ashore. In any event, he was learning basic navigation skills and applying them, along with leadership training.

Upon graduation in 2009, he was immediately commissioned as an officer and sent to a medium endurance cutter, the USCGC *Resolute*, in St. Petersburg, Florida. He recalled the awkwardness of the situation. Upon arrival at the ship, he was immediately given the position of second in charge of the Engineering Department responsible for thirty engineering personnel and upkeep of the ship's machinery and systems. Thus, at age twenty-two, he was in charge of the engineering enlisted people, at least one of whom was forty years old, almost twice his age. Of course, Jacobsen had to make decisions based on his rank, but the senior enlisted people would gently guide him along the way.

Looking back on his career, Torrey said that his tour of duty on the *Resolute* was an important learning experience. There, he started out in engineering. He learned how engines run and how maintenance is done. From time to time, he gained experience in ship driving, and he gained qualification as a deck watch officer. Since this ship was devoted primarily to law enforcement, he learned about basic Coast Guard missions such as drug and migrant interdiction as well as search and rescue in the Caribbean. After a couple of years on the *Resolute*, between 2009 and 2011, it was time to move on. For better or worse, the Coast Guard, like the other branches of the military, has a policy of frequent job rotation. There are some good arguments for this policy. It is designed to promote flexibility and cross-training for both officers and enlisted personnel. Partly, it is to keep them interested, but also it rotates them between high-stress assignments like sea duty and less stressful assignments like shore duty.

In any event, Torrey's next assignment in 2011 was on the USCGC *Juniper*, a seagoing buoy tender, stationed in Newport, Rhode Island. There his role was as operations department head, responsible for the mission execution of a 225-foot seagoing vessel and fourteen personnel. This assignment represented a change of tracks—from a law enforcement vessel to a working vessel—one that was dealing with the care and maintenance of the signs, symbols, buoys, markers, and regulations so that vessels can safely navigate the maritime environment. While assigned to the *Juniper*, he spent two weeks at the National Aids to Navigation School in Yorktown, Virginia, to receive a more formal education in the markers for the nation's waterways.

Beginning in 2013, he was assigned to the Office of Navigations Systems at Coast Guard Headquarters in Washington, D.C. This office is responsible at the headquarters level for management of policy involving all waterways of the United States, including 92,000 aids to navigation over 25,000 miles of navigable waterways and 350 ports. During our interview, I was inclined to skip over the headquarters chapter in Torrey's career. I was always more interested in the down-to-earth operational workings of the buoy tender in the New York–New Jersey Harbor.

However, Torrey politely and confidently corrected me. I came to understand the Coast Guard's management perspective on ATON is actually interesting, even fascinating. What I did not understand is that ATON is the most expensive mission in the Coast Guard. Why is it so expensive? Like the other missions, ATON requires the expense of ships and crews. However, there is also a great deal of costly infrastructure required. The more you look into the problem, the more you realize how costly this mission has become. For example, steel buoys become eroded and rusty from extended periods of time in salt water. After some six to nine years of service, a steel buoy needs to be transported to an industrial facility to be overhauled. There it is cleaned and repaired before painting. Once the process is complete, the buoy has be transported to its assigned position.[6]

Given the high cost of the ATON mission, I was not surprised to learn that there are some who advocate eliminating large parts of it to save money. This is the kind of issue that Coast Guard Headquarters must deal with. At first glance, the argument appears to be persuasive. It is true that large commercial vessels are equipped with the Electronic Chart Display and Information System (ECDIS). This is a computer-based alternative to paper charts, and it works really well. I have seen it myself many times, and it is amazing. Thus, a logical next step might be to simply place a mark, or a virtual buoy, on the ECDIS chart showing where the eliminated buoy used to be. This all sounds good until we stop to realize that sometimes computers fail and sometimes computers can be hacked by unauthorized users. Furthermore, we must realize that there are many boats without ECDIS including commercial fishing boats, domestic tow boats, and recreational boats.

Much the same arguments can be made for and against the GPS. Yes, most of the time it works very well. However, experts warn us against overreliance on GPS. For example, Dr. Brad Parkinson, regarded as the father of GPS, said, "Reliance on satellite navigation and timing systems has become a single point of failure for much of America and is our largest, unaddressed critical infrastructure problem. This is because GPS is a distant, faint signal that is very easy to disrupt. In fact, it is being actively disrupted every day. Fortunately, most of these disruptions are very local and of short duration. Occasionally, however, they can cause economic loss and can threaten safety of life."[7]

In the end, experienced mariners are always looking out the wheelhouse window. Looking out for navigational buoys enhances "situational awareness" and helps to prevent collisions and groundings. "Mariners rely on multiple layers of information to establish their positions, and the foundational layer that they depend on most is the physical objects they see out the window," said Captain Lynn Korwatch during a congressional subcommittee hearing. "In fact, many of the nautical charts specifically warn mariners not to rely solely on any one means of navigation."[8] To be sure, it would be nice to save the money being spent on ATON, but for the safety and security of our mariners the ATON mission must continue. Perhaps the best argument for funding comes from the everyday experience of driving our cars. Most of us rely on our GPS devices, but we still look out the windshield and see street signs, stop signs, road barriers, detour signs, and traffic lights.

After talking with Lieutenant Jacobsen, I came away convinced that the three years he had spent at Coast Guard Headquarters were not just a break from sea duty but a learning experience that gave him a deeper understanding of Coast Guard operation. In 2016, his next assignment was as executive officer for the USCGC *Elm*, a 225-foot seagoing buoy tender stationed in Atlantic Beach, North Carolina. This assignment represented a big increase in responsibility. As executive officer, he was second in command, reporting to the commanding officer. On a Coast Guard cutter, this meant that he was responsible for most day-to-day activities, freeing the commander to focus on the big picture. As we talked about his experience on the *Elm*, it became clear that the thing that left the most profound impression on him was having dealt with no fewer than three hurricanes.

As a New Jersey resident, I was familiar with the considerable onshore damage and flooding caused by hurricanes, but I had never considered the impact that hurricanes have on ATON. Torrey began to enlighten me: "Working buoys are very dangerous already, but I think most people do not realize that typically, after a hurricane passes by, the strength of the storm is incredible. It will lift those 18,000 pound sinkers and toss them like it is nothing, blocking channels such as the St. Mary's River down in Georgia going to King's Bay. There are vessels in there that are vital to our national security. If anything were to happen, those vessels would be required immediately to get in and get out for operations. So, when the storm blows through and

FIGURE 15 Seaman Jessica Smith, USCG, a buoy deck crewmember aboard the USCGC *Katherine Walker*, secures herself to a buoy in order to check the overall maintenance of it while under way in New York Harbor, March 8, 2016. (Photo credit: Petty Officer Third Class Frank Iannazzo-Simmons, USCG.)

pushes the buoys off station, sometimes it will block the channel. They have to be moved and they need to be moved now."⁹

So now I understood that a hurricane could move the buoys around and even block the channel; but I assumed that, once the storm had passed, that the work could easily be carried out. Wrong. Torrey explained, "So, when the hurricane passes through, it does not mean the seas are now suddenly calm. The seas are not all stirred up, and it takes a while to calm down, so you end up working buoys in heavy seas and heavy winds, and the risk is tremendous. You are pulling the ship alongside a buoy in heavy seas, and you are trying to have a steady course while lifting thousands and thousands of pounds into the air because there is some urgency. I have tried to work buoys in 8-foot and 10-foot seas, twice as tall as you and me. You would perhaps wonder why this is necessary. But this is part of the security of our nation, right? It is tough for a crew from the deck plate perspective."¹⁰

After serving as executive officer of the *Elm* for two years, Lieutenant Jacobsen was next given command of the 175-foot *Katherine Walker* in May 2018. Of course, it is an important career milestone for a Coast Guard officer to be placed in the role of commanding officer (CO) of a vessel. As such, he has the ultimate authority over the unit. It is a big deal. At the same time, along with that authority comes ultimate responsibility. Before meeting with Torrey, I already knew that there were some 400 aids to navigation in the Port of New York and New Jersey. I knew, in a general way, that buoy tenders throughout the Coast Guard followed a regular maintenance schedule so that major problems could be prevented before they occur. But I wanted to learn more about the details.

Torrey patiently explained that the Coast Guard maintains computerized records for every single buoy. He compared them to medical records. The service intervals depend on the environment of the buoy. Some buoys should be serviced every year, while others can be serviced every two or three years. The workload of buoy servicing is rather predictable; there is a routine maintenance sequence. What is far less predictable is the maintenance of the chains that connect buoys to their anchors. In the Coast Guard vocabulary, buoy anchors that are typically made of cast concrete are referred to as "sinkers" or "rocks." Torrey said, "What wears the most on a buoy is the chain. The buoy is constantly moving back and forth, right? Over time, there is gradual wear and tear from the sand and current, not to mention major storms and vessels passing by."[11] He went on to explain that, after a number of years, they will have a good idea of when the thickness of the chain links has become too small for the capacity of the buoy, and there is a real danger that it might break loose during a major storm. So, they go out and pull the whole thing up on deck, cut that chain out, and put a new chain connecting the buoy to the sinker.

On the one hand, it is true that much of the work of buoy maintenance is scheduled and routine. On the other hand, that does not mean that it is easy and risk-free. As Torrey explained, "It is incredibly dangerous work really, and you have to imagine that we are bringing a 900-long-ton ship right alongside a buoy that is marking a shipwreck or shallow water that everyone else is trying to avoid, and we are pulling up right alongside." As we continued this discussion, I argued that the *Katherine Walker* has a relatively shallow draft, which makes it easier to approach a dangerous area. Torrey agreed that the ship has a shallow draft and that it is very maneuverable, but still there are many risks in bringing a vessel alongside a shallow area. For example, there might be a rock that is sticking out of the water that no one can go over. A small wave might push the ship over that rock.

It seems that there are unavoidable risks in buoy maintenance. There are risks to the safety of the crew and risks to the ship itself. How do you minimize exposure to risk? Of course, risk cannot be eliminated, but Torrey

FIGURE 16 Crew members aboard the USCGC *Katherine Walker* work a buoy in New York Harbor, March 8, 2016. On the left is the safety supervisor. In the center is an experienced qualified rigger. On the right is a break-in who is still learning the ropes. (Photo credit: Petty Officer Third Class Frank Iannazzo-Simmons, USCG.)

explained that the Coast Guard has procedures in place to stay ahead of the problem. To protect the crew, training is key. Buoy handling is dangerous work because of the long hours, bad weather, and wet and muddy decks, as well as hand and finger hazards, not to mention the danger of getting in the way of a heavy buoy being swung aboard the ship. The crew on the deck is dependent on the skill and experience of the crane operator. Thus, a great deal of time is spent on individual qualifications, running people through boards and ensuring that they are held to high standards for each position onboard the ship. Performing the job requires months of dedicated training. Fortunately, the *Katherine Walker* services hundreds of buoys, so there is a great deal of repetition. The repetition allows the officers to qualify the crew members more quickly.

For the commanding officer, there is an inherent danger in the ship steering for a buoy tender. An experienced Coast Guard officer might have great confidence in basic ship driving, but this work involves a driving a ship into shallow waters, day after day. Formally, these shallow waters, full of dangers, are called "restricted waters." Everyone else is trying to give them a wide berth, but the buoy tender has to get up real close. An additional worry is all the maritime traffic—ferries, tug and barges, container ships—flying past the buoy tender, creating wakes. VTS tries to warn that passing traffic to slow down when passing a working buoy tender, especially in a narrow waterway like the Kill van Kull. Approaching a buoy in shallow water poses a challenge every time. The commanding officer has to use his best judgment. As Torrey explained to me, "In the end, it's just a buoy, right? So, the ship and the crew need to come first, and that is where your priorities lie."[12] I came away with a renewed appreciation for Lieutenant Jacobsen's crushing weight of responsibility.

I followed up by asking about the most important attribute of a commanding officer of a buoy tender. Torrey explained, "You need to understand that, even though you are the captain, you do not know everything. That is tough because you want to be this confident, knowledgeable leader—someone who is always going to look out for your crew, but I think humility is the word I am looking for. That is key because, when you do not know the answer—that is okay. You do not get lost in the fact that you are CO, and you outrank everyone. It forces you to take a perspective view of the work of the crew, and the fact that they do not work for you. What we are doing is very risky, and people have been killed doing it. When I am forced to give someone an order to do something that is going to be dangerous or maybe threaten their life or hurt them, I have a sense of grounding on that. So, to me, humility, I think is the biggest thing."[13]

As we were winding down our interview and I was thanking him for his help, Lieutenant Jacobsen reminded me that the ATON mission requires teamwork. To be sure, the *Katherine Walker* does the heavy lifting of big buoys in the New York–New Jersey Harbor, but that is not the whole story. There are many repairs and adjustments that can be done without dispatching an 800-ton vessel with a crew of twenty-four. Clearly, the cost per hour of operating the *Katherine Walker* is rather expensive when taking into account the cost of salaries, equipment, fuel, and so on. Fortunately, at the Coast Guard Station in Bayonne, New Jersey, there is not only the *Katherine Walker* but also a highly functioning Aids to Navigation Team (ANT). This is a shore-based unit with the mission of helping to maintain all kinds of ATON, with a variety of small boats, trucks, and other equipment. It is more cost-effective to send out a smaller boat with a smaller crew if they can take care of the problem. I decided to find out more about ANT.

Aids to Navigation Team (ANT), Bayonne

It turned out that the ANT for New York Harbor was originally stationed on Governors Island. However, as a cost-saving measure, in 1997 the ANT was moved to the Military Ocean Terminal in Bayonne, New Jersey, which served as an Army Base from 1967 to 1999. The long pier has since been renamed the Peninsula at Bayonne Harbor and is slated for mixed-use development for residential, industrial, commercial, and recreational uses. The ANT is located in a modern one-story building with white metal siding and a red metal roof, surrounded by a chain-link fence topped with barbed wire. It is a state-of-the-art facility with some 16,000 square feet housing offices and workshops. In front of the building is a large flagpole flying both the American and Coast Guard flags. Right across the street was a row of abandoned and boarded-up Army warehouses that were unsightly; however, I was told that it was simply too expensive to tear them down. Next door was a huge parking lot for imported cars brought in by car-carrying ships.

I wanted to learn more about this operation, so I introduced myself to the officer in charge, Master Chief Petty Officer Jason Willey. Tall and thin, with close-cropped red hair, he projected self-confidence, authority, and enthusiasm. He was quite willing to show me around the facility. Because he was friendly and accessible, I fell into the habit of addressing him as "Jason," but I noticed that the twenty-six regular active duty people in his crew always addressed him as "Master Chief," but never as "Sir," a term reserved for officers. As a master chief, he holds the ninth, and highest, enlisted rate, pay grade E-9, which places him in the top 1 percent of enlisted members of the maritime forces.

Jason wore the standard blue working uniform with blue pants and long, untucked shirt worn over a blue undershirt, and black work boots. The shirt had the words "U.S. COAST GUARD" in white letters over the left breast and his name "WILLEY" over the right breast. On both collars was the distinctive Coast Guard pin with shield and anchor, topped with two stars, indicating the rank of master chief.

Jason grew up in Morrisville, a small village in Lamoille County, Vermont, near Stowe, where he went to high school. Living in an inland, New England village, he had no maritime role models. He was bored with high school because he felt that he was not being challenged. The guidance counselor offered him a choice between taking the SAT or the Armed Services Vocational Aptitude Battery (ASVAB). Jason asked the guidance counselor, "Which one is easier?" He was told that the ASVAB was easier, so that is what he took. Later, he was rummaging around in a local thrift store and he came across a pile of secondhand military flags. The one that caught his attention was the Coast Guard flag. He said to himself, "That sounds interesting."

Shortly afterward, he signed the papers, and he was off to boot camp in Cape May, New Jersey.[14]

After boot camp, Jason was first assigned to the USCGC *Sorrel*, a 180-foot-long buoy tender, stationed in Staten Island, New York. It was a vessel dedicated to maintaining and replacing navigational buoys. He became fascinated with that kind of work, and he has stuck with it ever since. He still keeps in touch with several of his shipmates from that early period. Later, while holding the lowly rank of seaman, he met his wife, Erin, who was also serving in the Coast Guard as a seaman. In time, Erin resigned from the service to become a full-time mom. At the time I spoke with him, the couple had three children, two boys, thirteen and eleven, and a girl, six. I asked him if he would ever encourage his children to join the Coast Guard. He said, "Absolutely, because the Coast Guard is a small service with a strong sense of family. Besides, it offers great opportunities."

Later, I asked him if his Coast Guard career has been a positive experience. Jason responded, "Yes, definitely. When I came in, I was just an inexperienced young kid. The Coast Guard forced me to grow up quickly. It's tough love. You can't just call your mother if you hit a rough spot. You have to deal with it. What I learned is that the Coast Guard is not just a job, it's a lifestyle, and the quicker you realize that, the more successful you will be."

We talked about the problems of military leadership. Jason explained that he has to influence others to accomplish the mission by providing direction and motivation. He said, "You cannot be friends with your crew, yet you have to show them that you care about them. I do care about my people, but I show them that I care by making them follow the rules. Besides, I find that people want structure. I give them structure. I want them to come to work on time and to do their jobs."[15]

As the interview continued, I had many questions, but Jason seemed to take pleasure in explaining the work of his aids to navigation team. I learned that he was supervising the work of some twenty-eight enlisted personnel, without a single officer in sight. Typically, the unit would have six boatswain's mates and six machinery technicians, as well as a few electrician's mates, along with a number of unrated trainees. This team was routinely servicing small buoys, jetty lights, and lighthouses.

Jason explained that his unit had a remarkably high operational tempo. Much of the routine work involved responding to reports of discrepancies. There might be a structure discrepancy—something destroyed or damaged or leaning. A buoy might be missing, or off station, or sinking, or adrift. A lighted buoy might be extinguished or burning dim. Many discrepancy reports turned out to be mistaken, but all had to be investigated. On average, Jason said that the unit responded to four discrepancies every week, for some 208 discrepancies every year. At that time, 2012, there was the additional burden of moving some seventy-seven buoys around to accommodate the nation's largest dredging project in the Kill van Kull. These, unplanned, labor-intensive moves increased the workload.

Jason said that the key problem with a heavy operating tempo is that it becomes difficult to schedule ongoing training and regular drills. In the press of regular business, it is easy to overlook team building and mentoring. I was fortunate to visit the ANT in Bayonne on a training day, when regular work was suspended to review protocols and procedures. This was to be a day of both classroom review and practical instruction. Jason explained to me that such training days are essential because the average tour of duty at this unit is only four years. That means that every year there is a 25 percent turnover of personnel, hence the need for constant training and drills. The timing was lucky since nearly everything being discussed was new to me.

The Aids to Navigation Team in Bayonne has four small boats. These include two 49-foot boats with the unwieldy name of boat utility stern loading, or BUSL, pronounced "Bew-sull." Each is very stable and equipped with a large crane in the back capable of lifting nearly 5,000 pounds, but they are rather slow, with a top speed of 10 knots. This slow speed can be very frustrating. Like other Coast Guard workboats, the BUSLs have black hulls. Useful for serving harbor buoys, they are not seagoing vessels. They accommodate a crew of four, with the possibility of carrying three additional passengers, for a total of seven personnel. The BUSLs provide an interesting example of how the Coast Guard, perhaps more than any of the other armed services, gives tremendous responsibility even to very junior enlisted people. We might find a petty officer third class, a rank just above seaman, the lowest rank of a noncommissioned officer, in full command of a BUSL. The rank is the equivalent to a corporal in the U.S. Army or Marines. Here is a rather young sailor in charge of a $2 million boat, with responsibility for seven lives.

The unit also has two 26-foot craft with the awkward name of trailerable aids to navigation boat, or TANB, pronounced "tan-bee." Each one costs about $178,000. Powered by twin 150-horsepower outboard engines, these lightweight aluminum boats reach a top speed of 38 knots, though seatbelts are required for any speed over 30 knots. They carry a crew of four, with a maximum of seven passengers. There is seating for four persons—two up front for the coxswain and the navigator, and two right behind them. The boat has a roof and a windshield, but it is open on the sides, and it can be very cold in the winter. The team can take the boat by trailer to local ramps to serve a widespread variety of structures and buoys. They have a hoisting capacity of only 500 pounds, so they are not good for the heavy lifting of buoys. However, they are very useful for a quick fix of minor problems, such as replacing burned-out light bulbs or reprogramming LED lanterns. TANBs, being small boats, are not suitable for operations with 30-knot winds or 8-foot seas.

On this particular stand-down day, with regular duties put aside, certain sailors were assigned the role of "boat keeper" for each of the four boats. Each boat keeper was given a checklist of tasks to be done. The Coast Guard expects that each boat will have a ten-year lifespan. However, that target can be met only with proper care and maintenance. Jason told the sailors that, if they see something wrong, they should fill out a work order right away. Even something as minor as a broken windshield wiper should be fixed immediately. If there is oil in the bilge, it might cause a fire. If there is an accumulation of salt, it will corrode the battery. In short, the sailors were told to take pride in their boats.

The importance of frequent emergency drills was stressed. It seems that there had been a standing order issued by the master chief that, each time a boat goes out, the coxswain should conduct three drills. Apparently, this order had been overlooked. Understandably, in the press of day-to-day business, these drills can easily be overlooked. A particularly important drill is for a person in the water. A crew member might fall overboard for many reasons, such as a loss of footing due to a slippery deck or a sudden unexpected movement of the boat. Falling overboard is dangerous and life-threatening. It is important for boat crews to be prepared for such accidents. Another important drill takes place to prepare a boat crew for the possibility of striking a submerged object, such as a log. The crew should know the emergency procedures in such a case. This involves checking the boat for damage to the hull and the influx of water. Senior sailors were advised to use routine trips as opportunities to expose their boat crews to frequent drills. These might include drills for search patterns, reduced visibility, dewatering, and emergency engine restarting. Later, in private, one of the senior sailors was admonished not to treat the drills as a joke, to keep up appearances, and to model good behavior.

Finally, there was a review of the different types and sizes of buoys as well as how they are serviced and maintained. Here, I quickly felt that I was enrolled in a course that was over my head. There was a bewildering number of buoys, each with its own meaning and purpose. It turned out that buoys come in many different shapes and colors. They are often used to mark the sides of channels, obstructions, and anchorages. I came away with an amateur understanding of the basic types. To keep it simple, I learned that there are nun buoys and can buoys. On the one hand, a nun buoy is a conical-shaped buoy used to mark the right-hand side of a channel—that is, the right-hand side when facing inland. On the other hand, a can buoy is used to mark the left-hand side of the same channel. In other words, when a ship is going toward land, the nun buoy should be on the starboard side, the can buoy to the port. Of course, there was more to the lesson than such a simple distinction. The instructions covered buoys with lights, buoys with radar-reflecting material, buoys with bells, buoys with whistles, and so forth.

The buoys used to mark the channels in New York Harbor are anchored to the floor of the sea, as we have seen, by means of molded concrete called sinkers. The floating buoy is connected to the sinker by means of a heavy chain, each link of which is one inch in diameter. The chain comes in 90-foot lengths. Each such length is called a "shot" and weighs 1,000 pounds. In normal use, buoys must be swapped out every six years. On average, the chains are replaced every three years, depending on their rate of chafing. When approaching a buoy for servicing, the boat crew must unhook the chain from the swivel at the base of the buoy, being careful to hook the chain onto the boat, so as not to lose the chain. A refurbished buoy can be hooked back onto the chain, and the old original buoy can be brought ashore for treatment—usually involving sandblasting, patching, and repainting.

I came away with a deep respect for these men and women—their professionalism, pride, and enthusiasm for what they do. Especially because the work that they do is largely unnoticed. It is comparable to the work of a state highway commission that paints the stripes on the roadway and puts up the directional signage. Their work, also, is taken for granted. As Coast Guard missions go, aids to navigation work does not have the heart-rending glamor of search and rescue. It does not have the high visibility of law enforcement work. They almost never get to see their names in the newspaper. They just go about their business quietly and without fanfare.[16]

About a month later, I was invited back to Bayonne for a nighttime exercise in one of the TANBs. It seems that the Coast Guard small boat personnel are all required to put in ten hours of nighttime exercise every six months to maintain their proficiency. The plan for the exercise called for a patrol of the entire AOR (area of responsibility)—including the East River, the Hudson River, the Lower Bay, the Upper Bay, as well as Long Island Sound. Naturally, I jumped at the opportunity. Of course, I had toured New York Harbor in the daytime many times on outings with the Circle Line or with South Street Seaport, but never before with the Coast Guard at night.

It was October 16, 2012, in the early fall, and sunset took place at 6:14. The Coast Guard says that nighttime begins 30 minutes after sunset, so we planned to leave at 6:34. However, before we could set out, we had go through the Operational Risk Management Procedure. The idea here was to increase mission success while reducing the risk to personnel and resources to an acceptable level. The responsible officer considers everything including the weather and the sea state. The guiding principle is to accept no unnecessary risk, but to accept necessary risk when benefits outweigh costs. An obvious point on the risk assessment checklist was to make sure that everyone was wearing a life vest. In my case, it took a few minutes of rummaging around in the storage locker to find one that was extra large. Finally, the checklist completed, the decision was made to go ahead with the nighttime exercise.

As we pulled away from the small-boat dock, one of the crew members said, "It's time to let loose a FART." That struck me as rather rude, but he quickly explained that he was referring to a "Fast Aids-to-navigation Response Team."

It was a moonless and windless chilly fall evening. As we pulled into Upper New York Bay, the same crew member said, "It is F-A-C." Knowing the Coast Guard's love of acronyms, I took the bait asked, "What is a F-A-C?" Without missing a beat, he replied, "It means flat-assed calm." Indeed, the water was almost surreal, so very flat and so very calm. Of course, there was an irony in this observation. Neither of us knew then that in less than two weeks, on October 29, 2012, Hurricane Sandy would strike New York Harbor.

In some ways, Jason explained, it is better to go on patrol at nighttime. There is far less traffic because the recreational boaters are mostly safe at home in bed. It is even better in the wintertime, when no one goes out unless they must. Jason went on to explain that the best thing to do at night is to proceed slowly and with caution. Because you cannot see as well, it is wise to go slowly, especially in a narrow channel, so that you can avoid unlit obstacles such as floating timber. It is important to remember that a boat has no brakes—it cannot stop on a dime.

When boating at night, it quickly becomes apparent that a boat has no headlight. Turning on a headlight would be counterproductive because it would just produce a glare off the water. Not to mention that such a glare would destroy your night vision. In the same vein, I could not take notes in the dark, so I did my best to remember the experience. Obviously, the coxswain needs to have some backlighting to see his instruments and electronic charts that show buoys, jetties, exposed rocks, and docks. The trick is to minimize the backlighting of the charts to keep illumination to a minimum.[17]

In summing up the role of the Coast Guard's Aids to Navigation Team in New York Harbor, it is fair to say that the mission is to help mariners avoid the dangerous edges of the harbor—the shoals, the rocks, and the docks—with appropriate signage. The team places lateral markers on the edges of safe waters. All of this takes a great deal of effort and planning, but it is not enough. There is another layer of protection for our marine transportation system called the VTS, which we take up in our next section.

Vessel Traffic Service

On July 12, 2018, I had an appointment with Greg Hitchen, director of the Vessel Traffic Service (VTS) at Coast Guard Sector New York. Being a lifelong ship enthusiast, I had some idea of what to expect, but I wanted to learn more. I visited Hitchen at his office located at Historic Fort Wadsworth on Staten Island, New York. His office was in Tarden Hall, named for Captain James Tarden, the first captain of the port. It is a handsome brick building built in Georgian style. I felt very lucky to get a tour of the facilities that the public never sees. The staff here oversees maritime safety for the Port of New York and New Jersey—one of the largest ports in the United States.

Greg Hitchen grew up on the water in Massachusetts. He was familiar with the Coast Guard from a young age. Early on, he decided that he wanted to join. He got into the Coast Guard Academy, where he graduated in 1987 with a degree in electrical engineering. For the first twenty years of his career, he spent about half the time afloat on Coast Guard cutters. He was on five different ships, and he was in command of two them. His last command was the USCGC *Tahoma*, based in Portsmouth, New Hampshire. The cutter split its time between fisheries enforcement at Georges Bank and drug enforcement as well as migrant interdiction in the Caribbean. Hitchen said, "Unfortunately, it seems that I always spent the winters at Georges Bank and the summers in the Caribbean."[18]

Hitchen went on to explain that his staff tours, when he was not afloat, were in fisheries-related duties ashore. He said that he found the science and economics behind fisheries studies fascinating. Along the way, he ended up getting a master's degree in 1997 from the University of Rhode Island in maritime affairs, with an emphasis on fisheries enforcement. He retired in 2013 after twenty-six years with the rank of captain. He then had a civilian job for a couple of years before accepting the post of director of vessel traffic at Coast Guard Sector New York. I asked him about his job satisfaction in that role. He explained that he was happy to once again be part of the maritime community: "This is the next best thing to going to sea. I am proud to be part of an organization that gets these ships in and out of the harbor."[19]

Hitchen patiently explained how his operation works. They monitor the ships in the harbor using radar, closed-circuit television, VHF radiotelephones, and an automatic identification system (AIS) to provide navigational safety in a very crowded waterway. I had some familiarity with AIS from previous ship visits. It is a remarkable system. There are transponders on all ships over 300 gross tons that use radio beams to broadcast the name of the ship, its course, its speed, its dimensions, and its type. This information is picked up by VTS and by other surrounding ships. Together with radar, the AIS goes a long way to help everyone avoid collisions. Hitchen went on to explain that the Coast Guard keeps watch over the entire harbor through the use of some nineteen remote sites—with thirteen radar installations, twenty-six closed circuit television cameras, and three AIS towers. Meanwhile, they have ready voice communication with the ships by means of eight VHF FM transmitters.

VTS has often been compared with Air Traffic Control; and, on the surface, there are similarities. Both operations try to provide safety and to avoid collisions. However, the differences are very real. Airplanes move very fast, and air traffic controllers routinely issue direct orders to their pilots. Ships move slowly, and the process is very different. Hitchen explained, "We cannot gauge what the tides and currents are doing for a given ship at a given time. The maritime environment is very dynamic, and ships are less responsive to control than are airplanes."[20]

Thus, vessel traffic normally does not order the spacing of ships. They usually leave that up to the ship pilots involved. They prefer to have the pilots work that out among themselves. We find a good example at the narrow Kill van Kull between Staten Island to the south and Bayonne to the north. Many ships cannot meet in that area. In addition, they all want to transit that area at a specific tide time—slack water, the short period in a body of tidal water when the water is calm, and there is no movement either way in the tidal stream.

Hitchen explained that, as they begin to plan their day, on a twelve-hour shift, they would start to look for potential conflicts for ships that want to pass through a given area at the same time. They would then start to negotiate between the two ships as to who might go first. Vessel traffic normally does not give specific courses and speeds to its ships. When they give directions, it is usually in terms that are more general. They might say, "Do not proceed past this point" or "Do not meet this ship in this area." Then they let the ship pilots figure out to comply. It starts with a recommendation. It rarely gets to a direction, because normally the ship pilots go along with the recommendation. Interestingly enough, vessel traffic rarely talks to a vessel master on bigger ships because they are all required to take on pilots. Vessel traffic routinely deals only with pilots. Hitchen said, "A very experienced pilot knows the harbor like the back of his hand. It makes for a smooth evolution for us."[21]

I asked Hitchen about the rise in the number of small recreational boats in the harbor and the danger of collisions with big container ships. Hitchen explained that, indeed, there is a problem. In the first place, there are already many big ships in the harbor, and they all broadcast their position by means of AIS. Hitchen said, "You see more on your computer screen than you did before. It becomes a bit of a sorting issue."[22] On the other hand, the small craft and pleasure boats do not broadcast their position on AIS. Vessel traffic has to rely on cameras and radar for the smaller vessels. The casual weekend boater may not understand that it is difficult for a large ship to alter course or avoid colliding because that ship is constrained by its draft or is unable to slow or stop.

Hitchen said, "The small craft is gonna lose in a collision with a bigger ship."[23] Of course, vessel traffic is worried about that, but it is not as big a concern as two big ships that would collide. An unfortunate accident with a small craft will not close the port. However, if two container ships collide, they might have to close the port, which would mean stopping trade. Fortunately, there is a very robust education program for boaters in the harbor. Boaters are taught how to watch out for large ships and how to be careful. The Coast Guard Auxiliary and the active-duty boats do spend some time educating boaters. At certain times, they even keep the small boats out of the way of larger traffic if they are going through some of the smaller channels.

Knowing that the VTS relies heavily on AIS, I asked Hitchen about the problem of inaccurate AIS information. He agreed that sometimes AIS has the wrong information regarding the length or width of the vessel. This usually happens, when the original owners first listed the information, they got it wrong. So vessel traffic does quality checks on ships as they check in or check out. They have computer programs where they can verify the information. When they find mistakes, usually on smaller ships, they reach out to the captain of the ship and inform him that he is broadcasting the wrong information. He usually fixes it right away.

Hitchen went on to explain that the AIS broadcasts the antenna's location, not the ship's location. He said, "A 1200-foot ship is a lot of real estate for just one antenna. That's where radar is very helpful for us."[24] By overlaying the AIS information on a radar screen, they can immediately tell where the antenna is located, compared with the rest of the ship. The watch standers are particularly well informed about that. They also rely heavily on radar when a ship is anchored. They want to make sure that the anchor is not dragging and the ship is not drifting. They watch the radar return very carefully.

I asked Hitchen about the problem when large container ships run aground. He explained that it is difficult to predict a grounding for a larger ship because it takes up so much of the channel. Thus, a grounding could happen very quickly. What they do is to monitor the situation closely to make sure that the ship is operating correctly. If a ship has any kind of a propulsion loss, they send it to an anchorage. Fortunately, there are guidelines for tug-assist requirements for bigger ships—particularly in a constrained waterway. For example, even if a ship were to lose propulsion in the middle of the narrow Kill van Kull, the guidelines would dictate that the ship have three or four tugs alongside that will keep moving that ship until it can safely anchor or safely get to a pier, under the direction of a harbor pilot.

To enable me to better understand how things work, Hitchen arranged for me to visit the actual Vessel Traffic Center for the harbor. He took me to this highly secure area so that I might see the how the watch standers there monitor the movement of the vessels and how they provide navigational advice in real time. The large room is dimly lit on purpose to enable the watch standers to better focus on their computer screens. Some scientists even say that dim lighting makes people more rational, thus able to make better decisions. In any event, there are six people on duty per shift. There is a watch supervisor, an assistant watch operator, three radio operators, and one reliever.

Hitchen introduced me to Carrieann Dixon, watch supervisor, who patiently explained her role and showed me a couple of the problems that she was dealing with on that day. Dixon explained that she generally works a twelve-hour shift, starting at five thirty in the morning. The first thing that she must do is to fill out a plan for the day, which is constantly changing. Dixon examines the planned ship arrivals and departures, and then she looks for anything that could be a hazard or a risk to the port. She tries to mitigate that risk and make sure that everyone knows that there is a situation that could arise later on in the day. Of course, they try to make sure the situation never happens.

Dixon gave me a specific example of a problem that she was dealing with on that day, and she was able to show me the problem on her computer screen. It seems that there were two large container ships already docked on either side of the relatively narrow Elizabeth channel, across from each other. There was a large Evergreen container ship at 1,100 feet on the north side, and a large Maersk container ship at 1,205 feet on the south side. Later in the day, another large container ship was scheduled to enter the channel. It would be tight, but it would be possible to get by both ships. However, there was a problem. While docked for loading and unloading at port, container ships routinely need to refuel. They do this by bringing a fuel barge alongside. Now, the channel becomes even narrower, since a typical fuel barge might be 190 feet long and 46 feet wide. It is hard to plan because the fuel barges may come alongside at any time. The fueling may take as little as two hours or perhaps as much as twelve or fourteen hours.

Dixon wants to make sure that the new ship coming in or going out does not have to maneuver excessively—moving back and forth, left and right. It is safer if the ship just has a straight shot. How to solve the problem? Dixon can request that the fuel barge delay its arrival to allow the other ship to pass safely. Usually, a simple request does the job; however, if necessary, Dixon does have the authority to order a delay.[25]

Watch supervisors like Dixon spend much of their time on the radiotelephone with the Sandy Hook pilots, the licensed maritime pilots for the Port of New York and New Jersey. These pilots guide oceangoing vessels in and out of the harbor. As we have seen, the watch supervisors work closely with the pilots, but they seldom issue direct orders, generally deferring to the pilots on the ships who are dealing with the all the on-the-spot variables such as the wind and the currents and with visual sighting of surrounding vessels and channel markers. In our next chapter, we will take a closer look at the Sandy Hook pilots and the harbor pilots.

7

Pilotage

● ● ● ● ● ● ● ● ● ● ● ●

In our previous chapter, we presented the work done by the U.S. Coast Guard in safeguarding the marine transportation system. They maintain the buoys, the fog signals, and the day beacons. They maintain "the road signs on the waterways." They also operate the Vessel Traffic Service to help guide the ships in and out of the harbor. In this chapter, we explore the world of pilotage. The pilots do the actual work of ship steering, bringing the ships into the harbor, and safely docking them. In all of the large ports throughout the world, the use of pilots is mandatory.

To be clear, there are two basic types of pilots serving in the New York–New Jersey Harbor, with two different sets of skills. The first group are known as Sandy Hook pilots. They are the ones who bring the ships from the open ocean into the Lower New York Bay. For the numerous ships headed for Port Newark or Port Elizabeth, the Sandy Hook pilot hands off control to a "docking" pilot, normally at a point just north of the Verrazzano-Narrows Bridge. The docking pilot, with the aid of tugboats, carefully takes the ship through the channels and nudges it into a berth. In this chapter, we first present the work of the Sandy Hook pilots, followed by the work of the docking pilots.

For the Port of New York and New Jersey, all ships with a foreign flag coming in off the ocean must have a pilot when entering or leaving the port. These pilots belong to either the New York Sandy Hook Pilots or the New Jersey Sandy Hook Pilots; however, the duties and responsibilities are the same for both.[1] For both associations, the senior pilots select and train their apprentices and set their pay. The way it works is that the shipping lines pay the pilot associations. The charge for piloting is based on the draft of the ship. In other words, the deeper the ship rides in the water, the more expensive the service. Then the associations pay each pilot, in shares, set by his license rank. All of these pilots must be licensed by the U.S. Coast Guard and by one of the two states.[2] The work of these pilots is reviewed annually by either the New York or New Jersey commission.[3]

The base of operations for the pilots is one of two station boats. In the wintertime, it is the larger and more comfortable Pilot Boat 1, the 182-foot *New York*, the more seaworthy of the two. In the summertime, it is the 145-foot Pilot Boat 2, the *New Jersey*. One of these is always stationed at the beginning of the Ambrose Channel. Both pilot boats have a distinct look with black hulls, white superstructures, and tan smokestacks and masts. Their hulls are marked "PILOT" in large yellow letters. Here the pilots make themselves at home while waiting for a call from an incoming ship. Pilots are listed on a rotation board. The one whose name is at the top of the list will be sent to the incoming vessel on a power launch. As this is written, there are four 53-foot American class launches (*America*, *Wanderer*, *Phantom*, and *Yankee*) used for pilot boarding and transporting pilots to and from the pilot station at Edgewater Street on Staten Island.[4]

In modern times, we have seen the implementation of new electronic systems that make guiding a ship from the open ocean into port much more straightforward. We now have the Electronic Chart Display and Information System, a big improvement over paper charts because the new system displays real-time information on a screen. Along with this, we have the automatic identification system (AIS), a tracking system that uses transponders on ships to reveal both their identity and their location. That information shows up immediately on the electronic charts. Given all that information, one might think that we no longer need Sandy Hook pilots to guide our ships into port. However, experts tell us that the pilots are more needed now than ever before. One such expert, Nick Cutmore, explains, "We are trying to get a quart into a pint pot. We now have these mega vessels; you've got ships now plying the seas that are too big to get into US ports, which is quite significant. Pilots are called upon to deal with greatly reduced turning basin clearance, very little keel clearance, and it's expected. So pilots have to be better than they used to be."[5]

I had the good fortune to meet a number of Sandy Hook pilots and their apprentices at the annual NY/NJ Port Industry Day in September 2018 in Jersey City, New Jersey. Particularly helpful was senior pilot captain John

Oldmixon. In a number of subsequent interview, he shared with me his career and some of his most notable experiences. John Oldmixon II came from a family background that explains how he followed this career. His father was a Sandy Hook pilot, as were his grandfather and his great-grandfather. John told me modestly, "We go back a little ways."[6] John explained that, until about twenty years ago, most of the pilots were related to other pilots in some way—not necessarily as sons. He went on to explain that there was a reason for those relationships years ago. He readily acknowledged that those arrangements are no longer acceptable today because of concerns over nepotism. However, he insisted that, years ago, it was not easy to find people who wanted to do the job. Who today would be willing to work for five years as an apprentice at minimum wage to get the job? In the post–World War II period, few people wanted to go out on the ocean and start driving little motorboats.[7]

Maritime piloting is one of those few occupations in America where there has always been, at least until recently, a strong tendency toward tradition. For a son to follow in his father's occupation could be construed as keeping the faith. We are all familiar with the time-honored proverb "like father like son." There is something to be said for the positive feeling that a son has when he feels like a link in a family chain that reaches back in time. Perhaps, in this regard, maritime piloting can be compared to farming or police work or firefighting. We are all familiar with those heart-warming newspaper feature stories when the son of a police officer grows up and follows the same career path. Invariably, the son says something like, "I always knew that this is what I wanted to do." These stories grab our attention because we value the idea of respect for elders, and we value those who have a sense of duty.[8]

At the time of our interview, there were about seventy-two Sandy Hook pilots on the active-duty roster, about half the number of when Captain Oldmixon got started. I asked him, "Why so few?" He explained that because the ships are bigger, there are not as many of them. Today there are fewer ships, each with more cargo. The number of Sandy Hook pilots is a self-policing number. The association gets urges and nudges from the state commissions about whether there are too many or too few. The association has a governing committee, and every year they decide on how many or how few apprentices will be accepted. The committee tries to project out how many will be needed. In a typical year, they take two apprentices. They have many retirements coming up in the next few years, but they have made pilots to replace them already. The new pilots are already onboard and working. It is difficult for the association to predict future needs, but they feel that they are better at it than an outside entity would be.

Captain Oldmixon explained that the commissioners, years ago, told the association that they had too many pilots. There was pressure to reduce the number of pilots. John told me, "To some extent, they were right. However, once someone is a pilot, what are you going to tell him? You're done? You cannot really do that." They started to depend on attrition, and they got their numbers down. They have had just about the right number for the past twenty years. He told me that, at the present time, there is a good balance between supply and demand. It is true that, after World War II, there were too many pilots, most of whom had served during the war. During that war, there were thousands of ships coming in and going out of New York Harbor. On a typical Friday night, there might be more than one hundred ships departing New York—all in one night. At that time, people were needed.[9]

I asked Captain Oldmixon about the pattern of his work. He explained that the Sandy Hook pilots work four weeks on duty and two weeks off. For those who are on duty, there is a "working board," or staff duty roster, of some twenty-eight or thirty pilots that do all the work for that month. When a pilot reaches the top spot on that board, he gets the job. It might well be a night job. There is a great deal of night work because most ships move at night. They want to be at the dock in the early morning because the dockworkers start their working day at eight o'clock in the morning. Typically, the pilot will work three such jobs, and then he goes back to the bottom of the list. He may then go home and rest for twenty-four hours before returning to work. John went on to explain that a pilot should not do this job when he is tired: "If you somehow make a mistake when you're tired, you will be in trouble with the Coast Guard. Part of your job is sleeping. You have to be rested to go to work, just like an airplane pilot."[10]

The regular pilot boat stays on station and serves as a place of rest for the pilots and place of training for the apprentices. It is also, of course, a communications center for traffic control. When a pilot is called to a job, he is taken in a 53-foot aluminum launch that goes alongside the ship. The pilot uses a Jacob's ladder to get aboard the ship. I asked Captain Oldmixon if he ever worries about using the Jacob's ladder. He thought about it for a moment, and he gave an honest answer: "Well, yeah. I'm sixty-three now. I never used to worry, but I do now. I was always a very agile guy. I played a lot of sports, especially basketball. I had back surgery six months ago, which slowed me down a little bit, but it feels all right now."[11]

When boarding a ship that is in motion doing between 6 and 10 knots, depending on the weather, the pilot must climb the pilot ladder. Then the pilot gets up to the wheelhouse, on an unfamiliar ship, where he adjusts the speed and gives course and rudder commands in order to get to the dock on time. The pilot must think carefully about the situation and then be prepared to make a series of decisions. While doing this, he must project an air of calmness and authority. As Nick Cutmore explained, "You have to have a certain persona to do it. You are on your own, by and large; you have to carry the crew with a certain force of personality, and part of that is to have the innate competence and knowledge, but also to be able to carry the crew with you in this passage and be unflappable, even when you want to flap. It's not for everybody, and there are a number of entrants to the profession who get weeded out during the training phase because it's clear they don't have that persona."[12]

The pilot's job is all about safety. He is not to take any chances and jeopardize the ship just to get to the dock on time. Captain Oldmixon told me a story to illustrate the need to favor safety over speed. When he was a younger pilot, he was bringing a ship into the harbor and moving along. There was some urgency to reach the dock on time. He said, "Captain, if you are in a hurry, we can increase the speed." The older Greek captain said, "Pilot, I don't have enough time to be in a hurry." John paused to get my reaction. To tell the truth, I was confused. I did not understand the story, so John explained. It meant that, if something bad happens, all that time you tried to save, you lost.[13]

It is not a pleasant subject, but I felt obliged to ask Captain Oldmixon if he had ever had a collision. Indeed, he had such an experience, though clearly it was not his fault. He explained that he had been piloting a ship carrying scrap iron from the dock to the ocean. Such ships are notoriously old and difficult to handle. He recalled that it was on Valentine's Day, years ago. He was coming out by the Staten Island Ferry docks on the starboard side, when suddenly *he lost all steering*. There was another ship inbound, and they collided. There was a great deal of steel damage; but, fortunately, there was no oil spillage, and nobody was hurt.

Afterward, John had to go to two 8-hour depositions at a building in Lower Manhattan with twelve lawyers and himself. He said that it was unpleasant; and, hopefully, it will never happen again to anyone else. It kept him awake for a long, long time. He said that the Coast Guard was good to him and fair. He told them, "I'm a ship pilot. When I'm on a ship, I expect to be able to steer. If I can't steer, then what am I hitting?' The Coast Guard understood his point; however, the New Jersey State Pilot Commission was tougher on him.

In a very calm voice, John explained that they suspended him for ten days, and they required him to attend a ship-handling course at the Massachusetts Maritime Academy at Buzzards Bay, Massachusetts.[14]

As John described the hearing at the State Pilot Commission, I had a strong feeling of anger. How could they punish a pilot for something beyond his control? I asked John, "Weren't you angry?" He explained that a show of anger at the hearing was a luxury that he could not afford in front of an agency that had absolute control of his livelihood. Afterward, he and his lawyer went out for a drink together. John told me that, for the rest of his career, the memory of this incident haunted him.[15]

I asked Captain John Oldmixon if he ever had to replace an inexperienced helmsman. John explained that sometimes the confusion might be due to a language problem. One time, he even encountered a helmsman who was confused about left and right. Rather than insult the helmsman, John told the captain, "I think he is tired." The captain found another helmsman for the rest of the trip.[16]

I asked about another problem. I knew that reckless and inexperienced recreational boaters often trouble professional mariners in the busy and crowded New York Harbor. Unsurprisingly, John had a story of such an incident. He told of an occasion when he was out on an older cruise ship, proceeding from the Upper Bay of New York Harbor to the Lower Bay. There was a sailboat near the Statue of Liberty. The sailboat kept trying to cross in front of the cruise ship. John took evasive action. He had to go full astern. Then, he had to go full ahead and put the rudder over to left to avoid hitting the sailboat. The sailboat actually bounced off the bow of the cruise ship. John did everything possible to avoid hitting the sailboat. Luckily, neither vessel was damaged, nor was anyone injured.

John went out of the wing and took the registration number of the sailboat. John yelled at the sailboat operator, reprimanding him for recklessness. In turn, the sailboat operator yelled, "I have the right of way." Of course, this was a misinterpretation of the rules of the road. True, sailboats generally do have the right of way, but there are exceptions. In a narrow channel, a large motor vessel has the right of way because it is difficult for that vessel to maneuver. John was particularly shocked because there were children on the sailboat, and they were not even wearing lifejackets. John reported the sailboat operator to the Coast Guard, but he never followed up to find out what happened after that.[17]

By way of conclusion, I asked Captain John Oldmixon, if he would do it all over again. He answered that he would happily do it all over again, and then he went on to tell a story that illustrated the point. It was in 2005 that the apprenticeship period was lowered from seven and a half years to only five years. At this time John was serving as marine superintendent for the Sandy

Hook pilots. In this role, he was a manager who coordinated operations and maintenance for the vessels. As a manager, he was confronted in his office with a group of unhappy deputy pilots who had just completed their full seven and a half years: "This is not fair. These new guys got to cut short their apprenticeship, and they are going to be getting two and a half years more salary than us." John replied, "Well, I am forty-eight years old right now; and, if I had to start all over again, to do seven and a half years as an apprentice, I would do it." That argument gave the complaining pilots a bit of a pause, but later John did work a little something out. He adjusted the payment schedule for the unhappy pilots. The adjustment did not even out the inequity, but it lessened the blow.[18]

For many years, there was an amicable relationship between the Sandy Hook pilots and the docking pilots. Each group had its own area of responsibility, and things ran smoothly. Both pilot groups were required to hold Coast Guard licenses, but only the Sandy Hook pilots had state licenses, one from either New York or New Jersey. Then, on June 7, 1990, there was an accident that changed everything. A tankship, the BT *Nautilus*, ran aground while approaching the berth at the Coastal Oil Terminal in Bayonne, New Jersey, resulting in considerable oil spillage. Whose fault was this accident? A careful reading of the subsequent report revealed that the ship was only 20 feet out of the channel when it struck a rock. Perhaps the channel should have been marked with a buoy.[19] Nonetheless, the docking pilot was held responsible in the court of public opinion.

There had always been some in the Sandy Hook Pilots Association hierarchy who had wanted to take over the role of the docking pilots. Now, with this terrible oil spillage, they had a new argument. They said that the docking pilots were licensed only by the Coast Guard, not by the State of New Jersey. They said that it was a security issue, that only state-licensed pilots should be guiding ships in New Jersey waters. Many politicians got onboard with the argument, including the governor at the time, James Florio. The docking pilots were understandably upset because it seemed that the politicians were not recognizing that the bar pilots and the docking pilots had different skills. The docking pilots began to fear that their jobs would be absorbed by the Sandy Hook pilots. It was a valid fear because there were more Sandy Hook pilots, and they had more political influence.[20] A bitter political dispute began that lasted some fourteen years.

The dispute came to a boiling point in 2004. Spearheading the movement to erase the distinction between the docking pilots and the bar pilots was Tim Dacey, the chairman of the New Jersey pilotage board and a close ally of the governor at the time, Jim McGreevey. Another strong supporter of the controversial bill was Charles Wowkanech, New Jersey State AFL-CIO president. Also supporting the pilot bill was the Port Authority of New York and New Jersey. Things looked bleak for the docking pilots. On the other hand, the docking pilots did have a few powerful voices in their corner. The Coast Guard commander of the port, Captain C. E. Bone, did not buy the security argument, and he opposed the bill in a letter to the State Senate committee. Also supporting the docking pilots was Ed Kelly, executive director of the Maritime Association of New York, who was happy with the traditional division of duties. There was also a broad coalition of shipping companies and shoreside facilities in support of the docking pilots.[21] Fortunately, at the last minute a compromise was reached. The docking pilots agreed to becoming licensed by the state pilotage commission, and the Sandy Hook pilots backed off from trying to absorb docking duties. Today, there still may be some good-natured rivalry, but there is no longer a bitter battle.

As I was doing research for this chapter, I came across a problem. From time to time, maritime issues do make the newspapers and magazines of the New York–New Jersey area. Typically, news coverage happens with the arrival of a newsworthy vessel. In such accounts, the writers, more often than not, will give well-deserved credit to the Sandy Hook pilots for the role that they played in the safe passage of the vessel. However, there is seldom any mention of the role of either the docking pilots or the tugboat operators.

This pattern of news coverage can happen with the arrival of a celebrated cruise ship or of naval vessels during fleet week. As I was writing this chapter, it happened again with the arrival of the hospital ship USNS *Comfort* into New York Harbor for coronavirus response efforts on March 30, 2020. All of the newspaper accounts of that arrival that I read gave credit to the Sandy Hook pilots for the role that they played in the safe passage of the vessel. However, I failed to find a single mention of the role of either the docking pilots or the tugboat operators. It has largely been a neglected story, but I hope to shed light on it here.

The Role of Docking Pilots

When it came time to write about the job of the harbor pilot, I knew where to turn. I arranged to meet up with senior harbor pilot Simon Zorovich. I had known Captain Zorovich for several years, and he readily agreed to help

FIGURE 17 Harbor pilot Simon Zorovich leaves a McAllister tugboat to board the container ship M/V *Maersk Singapore* on January 30, 2020, to direct the docking of the ship, which was built in 2007 and sails under the flag of Singapore with a carrying capacity of 7,500 TEU. The ship is 1,099 feet long by 140 feet wide and is no longer considered a large ship because of the larger 1,200-foot ships coming in today. (Photo credit: Rev. David Rider.)

me understand his work. He works for McAllister Towing of New York. He was kind, generous, and outgoing during our interview in New Jersey. He spoke clearly and rapidly, so I was glad that I taped all of our interviews.

I asked Captain Zorovich to explain a typical docking assignment. When he is on duty, he typically is at home in New Jersey, ready to go to work as needed. It all starts with a phone call from the McAllister dispatcher. There is a ship inbound and going to Port Elizabeth. He might meet the ship just at the Narrows, after it passes underneath the Verrazzano-Narrows Bridge. First, he must get in his car and make a half-hour drive to the McAllister headquarters in Staten Island, near the Bayonne Bridge. There he will get on a tugboat, and it will take him about forty-five minutes to get to the Narrows, where he meets the inbound ship.

Captain Zorovich will come aboard the ship, and greet the captain of the ship: "You have to instill confidence in the captain that you can do this. Gray hair is helpful." Simon then addresses the Sandy Hook pilot. They will exchange pleasantries with each other. Simon explains, "We all know each other." Then they get to the important things such as the traffic that will be met as they approach the docks. They double-check the address of the berth that the ship is headed for. Then comes the formal transfer of authority:

HARBOR PILOT: What do you have her on?
SANDY HOOK PILOT: It's on slow ahead.
HARBOR PILOT: Okay, I got you. Thank you very much.

At this point, they may briefly talk about the weather, their families, or their vacation plans, but it is soon back to business.

Once the harbor pilot takes control of the ship, he is the only one issuing orders to the helmsman. Yet I had read that the captain is still legally responsible for his ship. I asked Captain Zorovich about this apparent paradox. He explained, "Yes, the captain is always responsible for his vessel. He can, at any time, decide that the pilot is not doing a good job. He can override the pilot. The captain can take the con, if he has a good reason—such as gross negligence on the part of the pilot. However, it has to be bad before it is wise for the captain to take over. Otherwise, if he refuses the advice of the local expert, and if anything happens, the captain will be in trouble. There are many good reasons why a captain would never do that. I have never had that happen. I never even heard of that happening. It is that rare."

I asked Captain Zorovich about controlling the speed of a ship. He explained that there is an engine order telegraph (EOT) on every ship. It is a communications system for the pilot on the bridge to request varying engine speeds:

- Full Ahead
- Half Ahead
- Slow Ahead
- Dead Slow Ahead
- Dead Slow Astern
- Slow Astern
- Half Astern
- Full Astern

Knowing that most docking maneuvers in the harbor are done at slow speeds, I asked him if it was difficult to steer at slow speeds. He replied: "Going dead slow at 5, 6, or 7 knots, you can steer very well, even at those slow speeds. For most ships, you can stop the engines and you are drifting for a while. With the momentum of the vessel, you can continue to steer below 5 knots, often

FIGURE 18   Captain Zorovich (standing) issues orders to the helmsman as he guides the M/V *Maersk Singapore* through the Kill van Kull as they approach the Bayonne Bridge with the master (seated), who is still legally responsible for the ship. (Photo credit: Rev. David Rider.)

down to 2 or 3 knots. Even then, if it is not steering any longer, you can do something. For example, if the ship is turning and you do not want it to turn any longer, if you want it to go the other way, you can give it a kick on the engine at dead slow ahead for a few seconds, and it will do what you want. You can actually steer down to very slow speeds. Then, as you are approaching the berth, the tugboats can assist."

I was curious about Simon's terms of employment. I asked him if he was an employee of McAllister Towing of New York. This is how he replied as the recorder ran: "We are employees in the sense that they pay us. They provide benefits like health insurance and pension plans. We are technically servants of the vessel. We do so under the auspices of the company. The company provides the tugs, and we provide the service. It is a good arrangement for both. The company knows who we are. They want experienced pilots because of their legal exposure. If I make a mistake, the shipping company could sue me and take my house or whatever I have. It would not even cover the paint job on a ship."

I asked Captain Zorovich about how he gets paid. He said, "We have an association. It is a pass-through operation for billing purposes. We have shares. The monies are put together. The bills are sent out to the shipping lines. They send a check back that goes into a pooled account. Then the money is split up in terms of shares proportionate to the pilot's level of experience. You start out at one-quarter share. At this point, you are given smaller ships to handle. After time and more experience, you may be voted into half-share status, then three-quarters share. Finally, after considerable experience, it becomes a full share."

Knowing that there are many variables in docking a ship, such as the wind, the tides, and the currents, I asked Simon how many tugs were needed in a docking. He said, "It depends on the size of the vessel. Typically, it takes two tugboats. However, as the ships have become bigger, it now becomes three tugboats. For the new 1,200-foot ships, it can be four tugboats."

Simon went on to explain, "By way of background, it should be pointed out that the tugboat rates in New York Harbor are very low because of the competition between McAllister and Moran. The tugboat rates in Europe are much higher. They are able to charge on the order of four times what we charge here."

I knew that working as a harbor pilot was not your typical nine-to-five job. I asked Captain Zorovich about his work schedule. He explained: "It is somewhat a matter of individual preference. We work it out with the dispatcher. For me, since I live close-by, I prefer four days on duty and four days off duty. However, some of the others live far away. For example, one of my colleagues commutes from upstate New York. For him, it is a five-hour drive to the port. Therefore, he prefers to work one week on and one week off. The same goes for another colleague who commutes from South Carolina. You have to remember that, when you are on call, it is a twenty-four-hour day. They can call you any time."

I went on to ask Captain Zorovich about daily fluctuations in traffic. He explained, "There certainly are both busy periods and slow periods. Container ship generally have to start operations at eight o'clock in the morning. That means the ship should be at the dock at seven. Thus, the ship will start coming in at Ambrose at four in the morning. It will then get to the Verrazzano at about five thirty, and it will finally arrive at the dock, as expected, at seven. It tapers off in the middle of the night. However, tankers are entirely different because they are less labor intensive. Time is less important, but sometimes you have to go at a certain stage of the tides. For Phillips 66 at Bayway near the Goethals Bridge, you have to be there at high tide, and that could be any time of day."

I asked him about seasonal fluctuations in traffic. He continued: "We get busy in the fall because the Christmas gift season starts in August. In the colder weather of November and December, you start getting heating oil. Fall is a busy period; but, after Christmas, it slows down a bit. In April and May, it is slower because the Christmas rush is over. The heating season is winding down. In the summer, traffic picks up again because people are taking vacations, and they need gasoline for their cars. That push drives the tankers. If the price of oil goes down, people are buying more of it and storing it. If the price of oil is on the way up, they will buy because they want to buy before it goes up even further."[22]

In the shipping world, they measure the volume of cargo in terms of a 20-foot equivalent unit (TEU). It is a shipping container whose internal dimensions measure about 20 feet long, 8 feet wide, and 8 feet tall. The first generation of container ships, in the era from 1956 to 1970, were modified bulk vessels that could transport up to 1,000 TEUs. Over time, as the container ships were purpose-built, their capacity kept going up. In the period from about 1988 to 2000, you would see container ships with up to 5,000 TEUs.[23] By 2006, we were seeing very large container ships (VLCS) with some 11,000 to 15,000 TEUs. Then by 2013, we were seeing ultra large container ships with some 18,000 to 21,000 TEUs.[24] There seems to be no end to the trend for bigger and bigger container ships. As I was writing this chapter, Hyundai Merchant Marine (HMM) announced the launching of the world's biggest container ship—the 24,000 TEU container ship HMM *Algeciras*, named for the port city in the south of Spain, the first of twelve such ships to be built.[25]

Like many others, I find these larger ships fascinating, so I asked Captain Zorovich what it is like to handle them. He said, "You have to pay attention more. You have to concentrate going ahead. You have to be aware of small movements of the vessel. In other words, you have to be where you want to be. You cannot let it deviate much. You have to be on top of the situation at all times. It is a bit more of a challenge. You are impacted more by the wind because you have many containers stacked high. The other side of the coin is that everybody knows that you are driving a big ship, and they tend to get out of the way. That makes it a little easier."

The Kill van Kull is a three-mile tidal strait between Bayonne, New Jersey, and Staten Island, New York. It connects Newark Bay with Upper New York Bay. It is about a thousand feet wide. Taking a ship through it has been compared to threading a needle. I asked Captain Zorovich about it. This is how he put it: "Piloting a large container ship, looking down to either side, you cannot see the tugs. You cannot even see the water. You can see ahead, but not down the sides. You can get the same effect for yourself in your house. If your nose is on the table, so close to the table, you cannot see down the sides. All you can see is the table."

I asked Captain Zorovich if there was ever the case that he had to wait for high tide to gain another four and a half feet of clearance going through the Kill van Kull. He said, "It happened to me not too long ago. I had a ship to pass through there. My office told me that the draft of the ship was 46 feet. I thought, well, that is okay. That will work. Then I took a close look at the ship. It looks deep. I said to the tug Captain, 'Can you check the draft?' It turned out to be 48.5 feet. I knew that there was 50 feet underneath in the channel, but it was low water at the time. Moreover, you need a minimum of two feet of clearance. If you have a ship that is 106 feet wide and it rolls one degree, you have an additional one foot of draft. You must have two feet of clearance. I was going to have less than a foot. This was a problem. The solution was to wait for the tide to rise to provide more clearance. Had I not caught the shipping company's miscommunication in the report of the ship's draft, it could have been a problem."

I asked Captain Zorovich if an apprentice has to be voted into the pilots association after completing his apprenticeship. He explained, "You have to be voted in by the members as a partner, though no initial financial investment is required as in some other ports. You have to have a license, and the members have to approve of your work. In effect, the applicant must wait for a retirement vacancy because there is only a certain amount of income. It is divided by the number of people cutting the pie. The dispatcher tries to spread the work out among the pilots on call. There are always two dispatchers. One works the twelve-hour day shift from 6:00 A.M. to 6:00 P.M. The other works the twelve-hour night shift from 6:00 P.M. to 6:00 A.M. It is an around-the-clock operation."

I remarked that, based on my readings, the number-one job of a docking pilot is to keep the ship off the bottom of the harbor. Simon agreed, saying that he has to provide a safe transit, going from point A to point B. His job is to get the ship from the Narrows in a certain amount of time—safely and without causing any problems. It is important to get to the dock on time. For example, if there are longshoremen waiting to unload the ship at eight o'clock, the docking pilot needs to be cognizant of the time. For this, he gets aboard the incoming ship at the Narrows at five thirty or six. There are time constraints. He tries to move he ship safely and efficiently without a great deal of stress on the ship's crew.

I asked Captain Zorovich what can be done if the ship gets stuck on the bottom. He explained that it depends on when the vessel went aground. The worst time to be stuck is at high tide, because the situation is only going to become worse. When the water drops, you are going to be in real trouble. At low tide, it is not so bad. The pilot waits around, and things will get better. He can also use tugs to push and pull to free the ship. It depends on the severity of the grounding. If the pilot's maneuvering rips the bottom open, and there is a spill, that is bad.

He gave me an example of a grounding situation: "I was once called upon to free a ship in Lower New York Harbor. I arrived on the scene onboard a tugboat. I knew that it was at low tide, so I decided to have a cup of coffee and just wait for higher water. You do not know what the ship is resting on. It could be sand, mud, or anything. If it is a rock, you really do not want to be pulling the ship and dragging it across a sharp edge. I told the captain, 'Let us wait a little bit. The tide is going to rise, and hopefully the ship will come right off the bottom.' And it did."[26]

I asked Captain Zorovich about balancing safety concerns with the client shipping company's desire to speed things up. He explained, "There is often very little one can do if a ship is late. We always have to go at a safe speed. Going faster increases risk, and it only saves a few minutes. I think today everyone understands that. Years ago there was always a problem with delays caused by fog. In New York Harbor, there were no firm rules about how bad the fog had to be before coming to a stop. In those early days, people had different comfort levels with how far they could see. Let us say that you had a ship that was in the fog, and you decided to wait until it cleared. You would say, 'We are not going anywhere. We are going to wait until it gets better.' Invariably, somebody else, with a higher risk tolerance or perhaps bad judgment, is going to start moving. Then the dispatcher might call the pilot and say, 'What's the visibility?' 'It's not getting any better.' 'Well, another ship just went by. What are you going to do?'"

Captain Zorovich explained that this exchange was a form of pressure. It was not heavy-duty pressure. The dispatcher was not saying that he must go. It was soft pressure, as if to say, "Somebody else is doing it, thus you should do it."

Zorovich recalled a clever move by a port executive to resolve the ambiguity caused by indecision about what to do in the event of fog. Some years ago, a refinery executive in New Jersey had a realization. If he had ships coming into his terminal, he did not want a huge public relations nightmare. He could just imagine the headline in the newspapers: "Ship Goes Aground at Refinery." He was determined to reduce the risk. He did not want any ships coming into his terminal if the visibility was less than one mile. In a single stroke, the executive solved the problem by giving harbor pilots an objective marker. Unless the pilot could see ahead for one mile, he was not to move. In time, this became a formal Coast Guard regulation.[27]

Thus far, we have reviewed the work of a docking pilot in terms of getting the ship from New York Upper Bay, through the Kill van Kull, underneath the Bayonne Bridge, past Bergen Point, into Newark Bay. Now, let us take a look at the last mile, as the ship approaches the dock at Port Elizabeth. Imagine, for a moment, that we have a ship laden with consumer goods that has crossed the ocean from a place halfway around the world. At this time, there is just one more task ahead—to get the ship tied up to the dock so that the longshoremen can begin the work of unloading the ship. Just writing about it, I find the task intimidating and stressful; but, for Captain Zorovich, it is all in a day's work.

The docking pilot must make his approach and survey the docking area. At Port Elizabeth, for example, the side of the dock facing Newark Bay is about a mile long. Officially, it has berthing spaces numbered from 76 to 92, using only even numbers. Depending on the length of the incoming ships, there is room for somewhere between six and eight berthing spots. It is a bit confusing, because the berthing spots are approximate, not exact. It is almost as if you were parking a car in a parking lot without any painted lines on the pavement. For the sake of our example, let us assume that the terminal has asked that the ship be docked at a particular numbered berth, say berth 84. Fortunately, there are yellow rectangles with black numbers, six feet high, painted on the pavement on the dock, easily seen looking down from the bridge of the ship. These are not the berth numbers, but rather measurements in feet, with every one hundred feet marked.

So, in our hypothetical example, the docking pilot receives orders to place the stern of the ship opposite a certain specific foot marker, the spot where berth 84 begins. He makes his approach slowly and carefully, taking into account the current and wind conditions. He takes his time, proceeding toward the dock using tugs to nudge the ship, making small adjustments, under his direction. The tugboats follow the orders of the docking pilot because there has to be only one person directing the operation. Meanwhile,

FIGURE 19 Captain Zorovich on the bridge wing of the M/V *Maersk Singapore* that extends beyond the pilothouse to the width of the ship, which allows him a full view to help in the delicate last-minute docking maneuvering at berth 61 of Port Newark Container Terminal. (Photo credit: Rev. David Rider.)

the captain of the ship must completely trust the docking pilot. He places his concern for his ship, his reputation, and his company into the hands of the docking pilot. It is an interesting dynamic. Because the docking pilot is doing many things at the same time, the captain may interject a modest suggestion every now and then, to assist, knowing full well that he is not in his element. Captain Zorovich has compared the work of tugboats nudging the ships to the work of a sheepdog herding sheep.[28]

The tugboats, as many as four in number for the bigger ships, are available to assist in the approach to the dock. The tugs can push the ship at the bow or at the stern or in the middle. For pushing, the tugs are equipped with a bow fender to gently push the ship. From the point of view of the harbor pilot, the tugs are most useful when pushing the ends of the ship. Pushing at the stern can be especially tricky because the stern is often sloping downward. If the ship is riding high in the water, there is a danger that the tug pushing at the stern could slide underneath the ship, with the danger of metal-to-metal contact. Of course, if the ship is fully loaded and riding deep in the water, this is not so much of a problem.

Besides using the tugs to push the ship, they can also be used to pull the ship. The tugs do this by pulling on the tug lines connected to bollards, or strong upright posts, on the ship. The custom and practice at all New York and New Jersey terminals is that such ropes are provided by the tugs, not by the ships—for reasons of safety and liability. It is very important for the tugboat crew that the lines be in good condition. If a line were to break and snap back, it could injure or kill a tugboat crew member. Tugboat crews do not want to rely on lines provided by a ship because such lines would be unknown to them in terms of origin and condition. They will know if their own line is still in good condition. As the ship pulls alongside the dock, the situation is still dynamic in terms of wind and tide; but the tugs can hold the ship in a stationary position holding it against the dock.

Now we are ready for the next step, tying up the ship to the dock. The captain and the docking pilot on the bridge work closely together to direct the tie-up. They have to work together because the docking pilot knows the dock, and the captain knows his ship and his crew. The relationship normally goes smoothly because it has all been worked out by custom and practice. The captain typically might say, "Which lines do you want out first?" And the docking pilot might reply, "The spring lines." The spring lines are employed to prevent the ship from moving either forward or backward. Normally, there are two spring lines led to the bow to keep the ship from moving forward alongside the dock, and two spring lines at the stern to keep the ship from moving aft alongside the dock. Then there are the additional lines, which keep the ship against the dock. There are normally four such head lines and four such stern lines. Common sense would tell us that these lines would be straight across from the ship to the dock, but that would not work because they would be too tight to accommodate the movement of the tides up and down. Therefore, the head lines and stern lines stretch out at about forty-five-degree angles.[29]

These mooring lines are provided by the ship. There is a mooring crew at the bow with four or five crewmen under the leadership of the first mate, with a two-way radio communicating with the bridge. There is another mooring crew at the stern of the ship with four or five crewmen under the supervision of the second mate, with a similar two-way radio. Meanwhile, both the captain and the docking pilot must leave the pilothouse at the center of the ship and walk out toward the side of the ship facing the dock by means of the bridge wing, which extends outward, allowing a full view of the side of the ship facing the dock. Now, they can clearly see the dock, some 75 feet below. On modern ships, there is usually a complete control panel on the

FIGURE 20 The McAllister yard at 3165 Richmond Terrace, Staten Island, New York, on July 9, 2020. When a harbor pilot has finished with the docking of a container ship, he can be taken back to this yard, where we also find a number of off-duty tugboats. (Photo credit: Rev. David Rider.)

bridge wing. The captain, in consultation with the docking pilot, will direct the two mooring crews.

After receiving an order to put lines out, the mooring crew will throw out a lightweight rope, or heaving line. At one end of that rope is a "monkey's fist," or a type of heavy knot resembling a monkey's paw. The weight makes it easier to throw over to the dockworkers. At the other end, the lightweight rope is tied to the actual, heavy-duty mooring line. The dockworkers pull, hand over hand, on the lightweight rope until they get the regular mooring line. Then they typically employ a pickup truck with a special hook on the front bumper. They hook the mooring line to the truck and back up toward the bollard. The mooring line has an eye (or a loop) that goes over the shoreside bollard. Then, on the ship, the crew can use powerful winches, either electric or hydraulic drums, resembling giant fishing reels, to tighten the ropes. The dock fenders, typically made out of heavy duty rubber, prevent the steel of the ship from laying against the concrete of the dock. The most modern container ships have a system of automatic adjustment of the tension on the mooring lines. On the one hand, if the tension is too great, the hydraulic drum will pay out a bit more line to reduce the tension. On the other hand, if the tension is too slack, the drum can pull in more line to increase the tension.

Once the ship is tied up, the docking pilot can walk down the gangway onto the dock. He might then radio one of the tugs saying, "I'll meet you at the stern of the ship." Conveniently enough the bow of the tug is about the same height as the dock, so that the docking pilot can simply step off the dock onto the tug. The tug can then either take him right away onto the next job or back to the tugboat yard in Staten Island if nothing is going on at the time.[30]

8

Tugboats

● ● ● ● ● ● ● ● ● ● ● ● ●

In our previous chapter, we discussed the key role of the docking pilot in bringing a large ship safely through the harbor and getting it tied up securely at the dock. Clearly that is an important job, but it is also clear that the docking pilot depends heavily on his supporting cast of tugboats. Thus, the docking pilot is to the tugboat operators much as the quarterback is to the rest of the team. So, in this chapter, I wanted to take a close look at the life of a tugboat operator. As it happened, I had a ready-made introduction to such a person—Daniel Zorovich, the son of our harbor pilot, Simon Zorovich. This certainly made things easier for me, but at the same time it gave me pause. Why are some occupations more likely than others to being passed on from father to son? It is an interesting question that cannot easily be brushed aside. Fortunately, others before me have also pondered that question.

How do children choose what to be when they grow up? Research has shown that they often follow in the path established by their fathers. We are all familiar with that dynastic trend in such high-visibility fields as acting and national politics.[1] Statistics show that working sons of working fathers are typically 2.7 times likely as the rest of the population to have the same job. This trend has been shown to be true for bartenders, lawyers, doctors, and many others. Unfortunately, that kind of statistical information on some jobs such as tugboat operators is not available because of low sample size. Nonetheless, we have plenty of anecdotal information. It seems safe to say that some jobs simply are more interesting, challenging, and rewarding than others.

According to Kim Weeden of the Sociology Department of Cornell University, "It's not just a matter of education or what your parents can buy—there's something about the occupations themselves."[2]

The story of the Zorovich family begins with Jakov "Jack" Zorovich, who was born in 1919 on the small island of Losinj in Croatia. For the men of that island, there was not much choice: they had to go to sea. Jakov followed in that tradition, learning the ropes in small cargo vessels engaged in coastal trade. These small ships, called motorsailers, were originally designed as sailboats, but they were later equipped with small, supplementary engines. They were about sixty feet in length and carried whatever cargo was available—sand, logs, finished lumber, whatever. During World War II, Jakov was drafted into the Italian Navy, where he served as a helmsman on the destroyer *Leone Pancaldo*, until it was bombed and sunk on April 30, 1943, near Tunisia during a troop transport mission. Fortunately, he survived the attack and was rescued.

After the war, life on the island of Losinj was difficult, due to economic hard times. The people of the island were also resentful of the authoritarian and repressive regime of Josip Bro Tito. A bright spot for Jakov was the birth of a son, Simon, in 1948. But by 1951 conditions were so bad that the family decided to leave. Plans had to be made carefully because officially no one was allowed to leave. The family carefully observed the movements of the police patrols and slipped away under the cover of darkness. In a small boat of about twenty-five feet in length, they made their way to Ancona, a seaport in central Italy. The family arrived with no papers and no passports. Right away, they were placed in a refugee camp, where they languished for three years.

In 1954 the family plotted their escape from the refugee camp. Life in the camp was somewhat permissive. The refugees could leave the camp by day, but they were supposed to return by nightfall. The family communicated with their family back in Croatia, who agreed to come pick them up by boat—a sixty-foot motorsailer used as a coastal trader. By this time Simon Zorovich was six years old, so he has a fairly clear memory of the trip. The family headed west, across the Atlantic, with a determination to reach the United States. They stopped off en route at Agadir, Morocco, and later at the Canary Islands for fuel and provisions. In mid-ocean the crankshaft broke, so they had to depend completely on the sails. As a result, the best speed they could make was about 3 knots. The trip across the Atlantic took 34 days, and with the entire trip lasting 101 days. They arrived at the Dominican Republic, where they got the engine repaired.

The family arrived in Miami with no papers and no passports. It was not an auspicious arrival, but the family was not without resources. Luckily, Jakov had an older brother, Samuel Zorovich, who had legally immigrated to the United States years earlier. By this time, Samuel was a well-established businessman, who retained a lawyer to handle the family's asylum application. In time, Jakov moved to New York, where he looked for maritime employment. By 1966 he found employment with McAllister Towing & Transportation to captain the company's yacht, used to entertain clients. By 1968, he secured a Coast Guard license as a tugboat captain.

About the same time, Jakov's son Simon was a junior studying aeronautical engineering at New York University. But in 1968 the employment prospects for aeronautical engineers were looking grim because NASA had been laying off people. Meanwhile, his father had gotten Simon a summer job as a deckhand on the McAllister yacht. Simon quickly figured out that he could make almost as much money as an unskilled deckhand on the company yacht as he could as an engineering graduate. He said to himself, "This isn't so bad." And he never looked back.

Simon started working on McAlister tugs in 1971 at age twenty-three as a deckhand, a job for which no license was required. By 1979 he got a license and earned the rank of mate, by 1981 the rank of captain. At no point did he ever attend a traditional maritime college; he simply worked his way up on the job. By 1983, he obtained his pilot's license, enabling him to guide ships through the berthing and unberthing process, dealing with dense maritime traffic in tight quarters, taking into account the wind and currents. Back in the 1980s, typically a tugboat captain might also serve as a docking pilot. The tugboat would come alongside an arriving ship, and the captain of the tugboat would climb aboard the ship, at which point the mate on the tugboat would take over the role of captain. Nowadays, as we have seen, docking pilots are in a separate formal category. They are delivered to ships in need of service by fully crewed tugboats.[3]

Why did Simon's son Daniel follow his father's footsteps to choose tugboats for a career? There is no way to know for sure how that came about. However, I have spoken with both father and son at great length, and I have some good ideas of how it happened. Daniel was an only child growing up in a stable family environment with regular family meals. Researchers tells us that such meals are a symbol of shared family life and a key part of child development. The dinner table is an essential place for the socialization of children, a source of exposure to family conversation. For example, Professor Paul Fieldhouse of the University of Manitoba wrote, "The shared meal is an opportunity not only to eat, but also to talk, to create and strengthen bonds of attachment and friendship, to teach and to learn."[4]

In addition to the family dinner table conversations, there were frequent onsite visits to the father's workplace. Dan's father started bringing him on tugboats when he was five. Dan told me, "What five-year-old kid is not going to love tugboats?" Sometimes the crew would let the little boy drive the tug when the way was clear. When his father was piloting, he would bring Dan along. His father would leave the boy on the tugboat, and the father would go onto the ship that needed docking. Then the boy would hang out on the tugboat and listen to stories of the crew. Dan would be climbing all over the place and stealing cookies. He recalled that it was a lot of fun.

Dan realized that he really wanted a career in tugboats for himself one time when his father Simon did a job. Simon turned around a large container ship, backing it into a dock at Port Elizabeth. It was rather tight. Dan remembered looking back after his father had docked the ship. The crew on the tugboat waited for Simon to get off the ship. Simon walked down the ship's gangway and onto the dock, so the tug's crew could pick him up and go back to the yard. So Simon is walking back to tug, having just docked this huge ship. Dan must have been eight years old at the time, and he recalled thinking, "My father has this little tiny head compared to this great big ship, and that head made this gigantic thing do whatever he wanted." Dan thought that was pretty cool and he said to himself, "I want to do that."

Dan knew what he wanted to do. Both his grandfather and his father had made successful careers as tugboat officers by "coming up the hawse pipe"— the pipe going through the bow section of the ship that the anchor chain passes through. It is a colorful metaphor meaning that they both began their careers as unlicensed merchant seamen and did not attend any regular maritime academy in order to obtain an officer's license. They climbed up the rank structure through on-the-job experience. When Dan of the third generation came along, coming up the hawse pipe could still be done, but it would take a long time. Today, there are distinct advantages to attending a maritime academy. The student there pays his tuition, attends classes, and passes the courses. Under Coast Guard regulations, he is then allowed to go ahead and take the examination for his license. If he had just tried to work his way up on the job, he would have had to put in more sea time before being allowed to take the examination. Dan wisely decided to attend a maritime academy in order to get his license more quickly, which allowed him to get ahead faster in his career.

After high school, at age seventeen, he enrolled in the State University of New York Maritime College, known as SUNY Maritime for short. The college is located within historic Fort Schuyler on the Throgs Neck Peninsula, where they East River meets Long Island Sound. The college offers a four-year bachelor of science degree that leads to a third mate's license of unlimited tonnage. But that was more than Dan needed. He signed up for the small vessel program that offered a two-year associate's degree with a course of study that would prepare him to take the test for the 1,600-ton mate's license. It is important to point out that the key certification for maritime employment is not the academic degree but the Coast Guard license. Let us consider an extremely hypothetical example, comparing a job applicant with a PhD in maritime studies with one who has a high school diploma. Who gets the better maritime job? The person with the better Coast Guard license, regardless of any academic degree.

Dan's feelings about that college experience are ambivalent. Although he admits that the education was helpful toward getting his license, he now regrets that he missed out on some of the things that a regular college can offer. He would have liked to have a few electives in fields such as history or philosophy or music. At the same time, he intensely disliked the military discipline at the school. He has unpleasant memories about the shoe shining and such. As a member of the Regiment of Cadets, he had to follow a strict lifestyle with elaborate rules and regulations. Meanwhile he recalled, "I'm on Facebook seeing all my friends at regular colleges partying with cute girls and having a good time. And there I was sitting with my shaved head in the company of my two male roommates." But Dan put up with the unpleasant aspects of a quasi-military maritime academy because he was focused on getting a career on tugboats used for docking in New York Harbor.

For anyone working in the tugboat industry, there will always be lingering doubts about the value of formal instruction at a maritime academy. Dan gave me an example. When he was at school, they gave a lesson on how a tugboat should go about moving a barge. The way a tug goes about making up to the barge varies. Is the barge fully loaded? Or is it nearly empty? Is it windy? Is the tide coming in or going out? At the academy, this lesson is taught on a chalkboard. Meanwhile, out in the harbor, there might be a young high school graduate learning the same lesson firsthand by doing it. It is pretty clear that this sort of thing is best learned on the job. Even so, to get ahead quickly it is best to attend college.[5]

By 2012, Daniel found himself with a deckhand job on the *Bruce A. McAllister*, a twin screw tug, rated at 4,000 horsepower.[6] The very first task for the tugboat was to tow a barge up to Canada. Fortunately there were two

FIGURE 21 Daniel Zorovich's first tugboat job was as a deckhand in 2012 on the *Bruce A. McAllister*, a twin screw tug, rated at 4,000 horsepower, built in 1974 by Main Iron Works of Houma, Louisiana. Photo taken on July 16, 2020, showing the tug pushing a ship to the dock. (Photo credit: Rev. David Rider.)

experienced deckhands on this tugboat, so one of them showed Daniel what to do. It was all "on-the-job training." The barge was towed by means of some 2,000 feet of wire, meaning that there was a certain amount of slack in the wire, but not so much as to drag on the bottom of the ocean. There must be a certain amount of slack because if the wire is too taut, it might break, with disastrous consequences.

After a year's experience as a deckhand, Daniel got a lucky break. It seems that there was a mate on another tug who was called up for three months' service in the U.S. Navy reserves, so they needed a replacement. In 2013 Daniel found himself undergoing training to steer onboard the *Ellen McAllister*, a tractor tug with 4,000 horsepower. The captain and the mate would sit beside him and instruct him on fine points of steering. There was never any textbook study or simulator work. Everything was on-the-job training. By 2014, Daniel began filling in from time to time as acting captain of the *Ellen McAllister*, though he was still classified as a mate. Then in September 2015, he left McAllister to take a job as a mate with Moran Towing.

He was assigned to the *Jonathan C. Moran*, with 6,000 horsepower. The tug was 97 feet in length overall and 38 feet in width, with an 18-foot draft

FIGURE 22 In 2013 Daniel Zorovich took training to steer on board the *Ellen McAllister*, a tractor tug with 4,000 horsepower, built in 1967 by Marinette Marine of Sturgeon Bay, Wisconsin, for the U.S. Navy. In 2001 she was acquired by McAllister Towing. Photo taken on July 30, 2019, showing the tug escorting a ship with the New York skyline faintly in the background. (Photo credit: Rev. David Rider.)

when fully loaded. It weighed 283 gross tons. It had two engines and two so-called Z-drive propellers, meaning that they could each rotate 360 degrees, allowing for quick changes in thrust direction, making the tug highly maneuverable. Tugs with such a mode of propulsion are sometimes referred to as "tractor tugs," and they are especially useful in ship docking maneuvers. I asked Daniel if the tug had an elevated pilot house, which would be useful in pushing barges. He told me, "Since all we do is ship work, it's better to be lower, so we can get underneath the flair of the bow. The pilot house is low. Everything is low." At the time of my interview, the *Jonathan C. Moran* was brand new. The company took delivery on the vessel in May 2016. Built at a cost of some $14 million, the boat had many desirable features for crew comfort. It was designed to run as quietly as possible. For example, normally the generators for electrical power are located in the engine room; however, on the *Jonathan C. Moran* they are located in a separate room called the lazarette. This meant that if the engines were off and someone wanted to work on the engines, it would be nice and quiet.[7]

Daniel stayed with Moran for some four years, and he enjoyed the work. Most of the senior positions at Moran were filled with long-term employees, happy with their jobs, who had little inclination to move elsewhere or retire. Thus, in October 2019, he returned to McAllister as the mate for the new *Ava McAllister*, a 100-foot-long tug with 6,772 horsepower.[8]

FIGURE 23 In October 2019 Daniel Zorovich was appointed as mate for the new *Ava McAllister*, a 100-foot-long tug with 6,772 horsepower, built in 2018 by Horizon Shipbuilding Incorporated of LaBatre, Alabama. Photo taken on October 4, 2019, showing the tug escorting a ship in New York Harbor. (Photo credit: Rev. David Rider.)

One of the more interesting things about tracing an occupation over three generations is exploring the question of how things have stayed the same and how things have changed. On the one hand, the hard work and the long hours have remained the same. On the other hand, one big difference is how meal preparation has changed. Back some forty years or more, when Daniel's grandfather was working, each tug was assigned a cook. Old-timers remember those days fondly. We can get some idea of what things were like back then by looking up old newspaper articles.

For example, we can turn to an article written back in 1983 describing the work of a celebrated cook named John Gilroy. He was known throughout the harbor as one of the best. He took pride in doing a good job, and deckhands were always asking management to be assigned to his tugboat. The newspaper article described a typical Gilroy lunch consisting of "a shrimp cocktail with his own sauce; chicken soup made fresh from a stock he prepared a day earlier; a mixed salad of lettuce, tomatoes, celery, green peppers, and sticks of cheddar cheese; a 15-inch rib roast of beef; steamed asparagus; steamed corn on the cob; stuffed potatoes; rye, pumpernickel and French breads; freshly baked cheesecake and coffee."[9]

I asked Daniel how things have changed since then. He explained that the position of cook was lost during the contract negotiations of 1988. Nowadays, the four-man crew takes turns serving as the cook. I asked Daniel Zorovich about how things work today, when the crew has to do their own food shopping and preparation. He patiently explained that, with regard to food onboard the tug, the company pays for the food, following a formula of $22 per day per person. For four people for fourteen days, that works out to $1,232. On the first day, at the start of a two-week shift, typically the mate will go the office, get the cash, and sign a receipt. Then he returns to the boat and takes inventory to see what is there from the previous crew. There is always some food already there. The crew gets together in the galley and roughs out a menu plan that might include, say, meat loaf and lamb shanks and steak. They make a shopping list, like a regular household. Then they add individual favorite items. The mate might want diet ginger ale. The deckhand might want Fruity Pebbles for breakfast. The captain might ask for Pepsi as well as salt and vinegar potato chips. They phone the dispatcher and ask, "Do we have time to go shopping?" If the answer is yes, someone will volunteer to go to the store. During the two-week tour of duty, everyone pitches in with the cooking. Of course, there is only one galley. It is not like on a big ship with separate dining areas for officers and crew. As a result, the atmosphere is more family-like rather than military-like.[10]

Another significant change over the years has been a dramatic increase in the government regulations of tugboats. They now have no cooks but plenty of regulations. Traditionally, tugboats in the United States were uninspected vessels, but that changed in 2016 when the U.S. Coast Guard published a new set of regulations. Some in the industry have complained about the regulations being overly restrictive, but most seem to think that there are good reasons for these regulations. The new regulations are being phased in over time; but, in the future, each tugboat must have a certificate of inspection issued by the Coast Guard. The regulations deal with such things as the safe operation of the vessel, the presence of necessary lifesaving equipment in the event of a disaster, the presence of fire suppression equipment onboard, and minimum requirements for machinery and electrical systems.[11]

There are about fifty tugs that operate in New York Harbor. Most of them are used to push or pull barges, but a few help dock larger ships. The arrival of a large ship approaching New York Harbor sets in motion a choreographed chain of events. The incoming ship might be a container ship, tanker, bulker, or car ship. Typically, Sandy Hook pilots are responsible for guiding that ship from the entrance to Ambrose Channel to a point just past the Verrazzano-Narrows Bridge. Meanwhile, one of the tugboat companies has been notified to meet the incoming ship at that point. It might take a given tugboat forty-five minutes to sail from its home yard to the transfer point, where a docking pilot takes over from the Sandy Hook pilot. Then it might take another hour and a half for the tug to accompany the large ship to the exact spot where it needs to be carefully guided into a dock. For tugboat captains, the ride out to the ship and then proceeding on to the dock is the boring part. Where it gets interesting is when the tugboat, following the instructions of the docking pilot, eases the large ship into its berth.

Most tugboats used for docking larger ships today are equipped with what is called a Z-drive, a special kind of marine propulsion. These tugboats typically have twin screws, each of which can rotate 360 degrees. This ability permits rapid changes in thrust directions, hence rapidly changing the direction of vessel. Understandably, these tugboats are highly maneuverable, nimble, and efficient. The captain of such a tugboat must be good at multitasking. His left hand controls the direction and the speed of the port propeller; his right hand, the direction and speed of the starboard propeller. In other words, there are four inputs (engine rpm and angle of thrust for each of the two propellers) that he controls with two hands. Thus the boat can readily go forward, backward, or sideways.[12]

At the same time, there is more to worry about. The captain has to control the winch on the bow that controls the line attached to the large ship. He can adjust it with a joystick to pay more line out or pull more line in. Meanwhile, the captain is receiving orders over the radio from the docking pilot on the ship and communicating with the deckhand by means of a hailer, a loudspeaker with a built-in amplifier and microphone. Basically, the tugboat captain is doing four or five things at once during the docking process.

Docking tugs in the harbor at this time have a small crew, with only four people. At the entry level there is a deckhand. It is a job with a great deal of manual labor. During ship docking operations, the deckhand is a line handler. There is always a danger of injury in line handling, so it must be done carefully. During off-peak times, a deckhand may be required to do chipping and painting of the exterior. With regard to the interior, the deckhand is responsible for cleaning the bathroom and the galley as well as the common areas. When the tug is under way, he is supposed to check on the engine room at least once an hour.

The engineer is responsible for maintaining the engine as well as the deck machinery including pumps and winches. He is also responsible for the heating and ventilation systems as well as the electrical, sanitary, and refrigeration systems. With such a small crew, he is also expected to help with line handling during docking and undocking operations.

The mate is the second in command of the tug. He is typically on watch from noon to six o'clock in the evening and again from midnight until six in the morning. During that time, if there are docking operations to be performed, he functions much as the captain, steering the vessel as needed. When there are no docking operations, he often conducts various required drills such as man overboard and fire drills. After each drill, he must complete the required paperwork. He is also responsible for updating the charts—both the electronic and paper versions.

The captain, of course, has overall responsibility for the tug. He is typically on watch from six in the morning to noon and again from six in the evening to midnight. During docking operations, he signals the workers on the deck with instructions for line handling. While steering the tug, the captain must avoid hazards to shipping, using navigational devices, including radar and depth finder, as well as lighthouses and buoys. It goes without saying that the captain is the boss. He issues orders to his subordinates; however, this is the merchant marine, not the navy, so often the orders are disguised as requests: "Hey, man when you get a chance, would you please clean up the galley?"

Of course, the captain cannot be watching everybody all the time. Everyone has his own job and does it on his own. The captain cannot be watching the mate because the captain has to sleep while the mate is on duty. The mate cannot always be watching the deckhand. For example, if the mate is steering the boat, the deckhand will have to go about his duties unsupervised. The crew members have to rely on and trust each other. If a deckhand cannot do his job without being watched, he is either new or incompetent. If he is new, of course, the others will have to provide instruction and training. But once he is trained, he will be expected to do his job without supervision.

The tugboat crew members are not just working together but living together. Hence, good manners are important. The situation is comparable to four men living together. Everyone should do his part to take care of the boat. For example, if someone is making a sandwich, he should not just leave the bread and cold cuts out on the counter. These items should be put away. Dirty plates should be rinsed off and put in the dishwasher. It is not good manners to leave an empty coffee cup in the wheelhouse. Apart from cleanliness, there is the matter of punctuality. The schedule calls for six hours on duty and six hours off. The unwritten rule is that everyone should show up for work fifteen minutes before the official duty period begins. Most of these things are common sense.

There is a widespread feeling among old-timers in the tugboat fleet that the young men coming out of the maritime academies tend to be arrogant greenhorns. They come in with the attitude that they know it all. They think that they are better than the workers without licenses. They can be quickly disabused of this notion by the old-timers. Only six years out of the maritime college, Dan already identified with the old-timers. At one point, Dan had a twenty-one-year-old recent maritime college graduate on his boat as a deckhand. Everyone must start as a deckhand, license or not. Dan wanted to be welcoming and helpful. He did not want to be rude, but he felt that he had a professional duty to point out that the recent graduate still had much to learn. The new man already had one week of experience on another tug. Their initial conversation went like this:

> "If you have any questions about the boat or the work or the harbor, feel free to ask. There's no such thing as a stupid question."
> "Oh, I'm pretty good with the harbor."
> "Really? What's that dock over there?"
> "Ah... that's New Jersey."
> "You got the state right. Please continue."
> "Is that Bayonne?"
> "You're getting warmer; but, after a week, I wouldn't say that you are pretty good with the harbor."

Dan went on to explain that there are people who spend their whole lives in New York Harbor, and they never get to every nook and cranny of it.[13]

On a small tugboat with a crew of only four people, it is important that everyone gets along with each other and work in a harmonious and cooperative spirit. Having said that, there is a chain of command. Some people are empowered to give orders, and others are required to carry them out. To be sure, in the merchant marine as opposed to the U.S. Navy, orders are often disguised as requests or suggestions, but they still need to be carried out.[14]

The job of a tugboat officer has changed a good deal in recent years. With more regulations, there is more paperwork. In the old days, officers simply had to maintain the ship's logbook. This would be a record of basic information including the time of routine events and notes on significant incidents.

Now records must be kept on lifesaving equipment, noting its condition and any expiration dates. There are records to be kept on the automated external defibrillator (AED) and the emergency position-indicating radio beacon (EPIRB). Records must be maintained for the oxygen bottles and the fire extinguishers and the emergency lights with their batteries. There is even paperwork to record the details for the change of watch. Every company has its own lawyers, and the lawyers want to make sure that the paperwork is in order.

Although the compensation for a tugboat worker is good, there are problems with the job. The chief complaint is the schedule. Tugboat workers spend two weeks on the boat, twenty-four hours a day. Each day they are on duty for twelve hours (in two six-hour shifts) and off duty for the other twelve. Thus, they work 182 days a year and have off 182 days. As a result they miss half of everything. Depending on the vagaries of the calendar, in a given year they might miss Christmas and Thanksgiving, not to mention weddings, funerals, and birthdays. They have a great deal of time off, but no control over when that time off takes place. While they are on duty, the docking tugboat workers are on call. They wait for the dispatcher to send them to help an arriving ship into the dock or a departing ship out of the dock. Of course they are paid for the day whether the work is busy or slow. But most tugboat workers prefer to be busy while on duty. It is frustrating for them to sit all day in the yard, waiting for an assignment. Time goes by very slowly. Meanwhile, they may be thinking of all the activities they are missing at home. Being away from home for half the year can be costly for relationships and marriages.

Since tugboat operators are paid by the day, it is not easy to take a day off. Besides, if they take a day off, they have to find relief. Then they are in debt to that relief. It is best just to work the assigned days. On their days off, it does happen sometimes that they might get a call from the company, asking them to come in and work for someone who is sick or injured. They do not have to accept the assignment. Of course, if they do accept the assignment, they are paid for those extra days.

I asked Daniel about job satisfaction. He told me that when he is feeling idealistic, he sees his job as oiling the wheels of commerce: "I am doing my part to contribute to society when I am helping to get gasoline into the harbor so people can drive their cars. I am helping families when I am bringing in toys at Christmastime. I am helping college students when I am bringing in a boatload of computers." He went on to explain, "Job satisfaction on the tugboat is very different from job satisfaction in my hobby, which is singing. When I do a good job at singing, I get applause from the audience. If I do a good job on the tugboat, nobody comes around and claps. It's all about pride. I always give 100 percent. I want to impress my peers." He gave an interesting example. At one point there were two identical tugs attending a ship docking. The docking pilot asked the other captain to carry out a rather tricky maneuver. The other captain begged off; he did not want to do it. Now the docking pilot asked Daniel to do it. Not realizing that the other captain had declined, Daniel went ahead and said, "Okay, I will give it a try." It was very satisfying when it all worked out. Was showing up the other captain part of the satisfaction? It was unintentional, but yes, for sure.[15]

I asked Daniel to tell me about a time when he dealt calmly and effectively with a high-stress situation. He paused for a moment to reflect on the question. He said that there had been many such moments, but the trickiest was when the whole world was watching, a time when there was a media spotlight on the situation. He recalled that the USS *New York* (LPD-21), an amphibious warfare ship, had arrived in New York Harbor on November 8, 2013, for Veterans Week, with a great deal of fanfare. The ship was moored at Pier 88 on Manhattan's west side. This is what is called a "finger pier," one that is perpendicular to the shoreline, making docking and undocking trickier than dealing with a pier that is parallel to the shoreline. During the visit, sailors and marines from the ship took part in variety of well-publicized activities. The uniformed service members met with local students, and they cross-trained with New York firefighters and police. They attended a number of Veterans Week remembrances, and they marched in the scheduled parade. When it came time for undocking, under the watchful eye of the media, one of the tugs on the job was the *Robert E. McAllister*, with Daniel Zorovich at the helm. The ship was at Pier 88, with the bow in, so it had to be backed out. It was a difficult job for several reasons. First of all, the USS *New York*, like most Navy ships, had no bow thrusters, making the job a bit more challenging. Second of all, whenever one is backing a ship, the ship has no steering because the rudder is behind the propeller and is of no use. The steering has to be done by the tug at the tip of the bow, gently pushing the big ship one way and the other.

The *Robert E. McAllister* was the tug at the bow. Daniel Zorovich was told by the pilot over the radio, "I want you to push the ship out of the slip quickly and carefully. And, by the way, I was just talking to the captain, and he says that you should avoid hitting the highly sensitive and fragile underwater sonar device at the bow." Of course, he could not see the sonar device because it was underwater. So it was a delicate and stressful situation, with dozens of news cameras focused on the ship's departure. On board the tug, just to add to the stress, were four sailors with machine guns and a bomb-sniffing dog. Daniel said to himself, "If I mess this up, it's going to be on the evening news, and my career is finished." Fortunately, it all worked out. The warship was pushed out into the Hudson River, turned around, and headed out to sea, with its sonar gear intact.[16]

In terms of a long-range goal, Daniel envisions someday becoming a docking pilot. To do so, he would have to ride along with an established docking pilot for a number of trips on his own time—not during the two weeks he would be on duty as a tugboat captain. He would also have to pass a very detailed examination. It's a long and arduous process, but in the end he could look forward to a satisfying career.

There are many steps in the process for those who desire to become a harbor pilot. The first step is to get a federal pilot license for the relevant harbor. The applicant has to meet certain prerequisites prior to applying and sitting for various tests administered by the Coast Guard. In the case of New York Harbor, there are some fourteen sections that must be mastered, each requiring a separate test. After making at least twelve documented trips over a given section of the harbor, the prospective candidate can apply to take the test for that particular section. The candidate makes an appointment at the U.S. Coast Guard Regional Examination Center at Battery Park in Lower Manhattan.

The candidate must bring in a number of outline charts on large sheets of paper for the particular section being tested. The outline charts show the surrounding bodies of land, but the harbor portion is blank. Upon arriving at the center, the candidate's blank charts are inspected by an official to make sure that there are no secret hidden marks on the paper. The candidate then must draw in, by hand, all the relevant details. For example, the candidate must draw in all the buoys in that section, identifying them by characteristics such as shape and color. In the Upper Bay section of the harbor, for example, there are more than one hundred buoys that must be included in the drawing. The drawing must accurately reflect the sides of the channel, the junction of one channel with another, obstructions, anchorages, and so forth. The candidate could easily need seven hours to accurately draw the chart. It takes about two weeks of intense studying prior to taking each test. Most tugboat workers will use a two-week period when they are off duty to do the studying.

After passing all fourteen test sections, the candidate will have a Coast Guard–issued license as a first class pilot; however, holding a license is no guarantee of employment. The next step is to seek endorsement from the New Jersey Maritime Pilot and Docking Commission to become an apprentice pilot. With this endorsement, the candidate can begin a rigorous on-the-job training program. During that time, the candidate will make some 250 additional rides with veteran pilots, observing and handling—under instruction—many ships.

After completing all this, the candidate can apply to the local pilots association for membership. The established pilots take a vote on the prospective candidate. They are very cautious about taking on a new member because the salaries are pooled. For example, in a given pilot association, there may be sixteen pilots. On the one hand, they do not want too many members because an increase in members means a decrease in each member's share. On the other hand, if business is brisk and the members are working long hours, they may welcome a new member to help share the workload. It is also possible that the established pilots approve of the prospective member, but they decide to tell the prospect that he will have to wait for a period of time until one of the active members decides to retire.[17]

9

The Contemporary Port

● ● ● ● ● ● ● ● ● ● ● ● ●

The port of today has come a long way since Malcom McLean sent that first container ship, the *Ideal X*, from Newark to Houston in 1956. Since that time, the original Port Newark was expanded to embrace a huge chunk of real estate in nearby Elizabeth, so that we can properly speak of Port Newark–Elizabeth. The port has no fewer than three major container terminals that are served by an on-dock rail facility, serving acres of warehouse space and many terminal buildings. In any given year, the port handles millions of tons of cargo. Arguably, the most important change for our port since the 1950s is that the container ships have been getting bigger and bigger. The reason for this increase in size is obvious: the bigger the ship, the more containers can be carried, and the lower the cost per container.

To properly tell the story of bigger ships and their impact on the port requires a quick review of recent history. Our story begins with federal legislation in 2000 that authorized dredging and deepening the main navigation channels in the Port of New York and New Jersey that support the container terminals. The project began construction in 2004 with the goal of providing fifty-foot depths all the way from the Atlantic Ocean to the Newark Bay Channel.[1]

The U.S. Army Corps of Engineers deepened the thirty-eight miles of channels in stages. The easy part was removing sediments and debris from

FIGURE 24 Dredging in Port Elizabeth on June 15, 2011. The project involved dredging thousands of cubic yards of heavy metal and petroleum-contaminated sediments that were excavated using a cable arm and a clamshell-style bucket. (Photo credit: Richard Levine / Alamy Stock Photo.)

the bottom of the channel. More challenging was the work of blasting and disposal of the granite bedrock. The work took longer than expected because there were years of delays in litigation pitting the Port Authority against environmental groups.[2]

Our story continues with the Republic of Panama announcing an expansion of the Panama Canal back in 2006. Almost immediately, the Port Authority and its maritime partners realized that a major change would have to be made at the Port of New York and New Jersey to accommodate bigger ships. The Bayonne Bridge spanning the Kill van Kull waterway would have to be raised from its 151-foot clearance to a full 215-foot clearance to allow the world's largest ships to pass. All of this work would require a great deal of planning, interagency cooperation, political support, and engineering expertise. To be sure, there were bumps along the way. In today's society, one cannot just push ahead with ambitious engineering projects; one must file elaborate environmental impact statements. Amazingly, it all worked out. Let us briefly review the story.

Panamanian president Martín Torrijos proposed an expansion of the canal in April 2006. A national referendum on the proposal was held on

October 22, 2006, in which Panamanian citizens approved $5.25 billion for the expansion. The referendum passed with 76.8 percent of the vote.

Deepening and widening of the canal began in earnest in 2007.[3] The plan called for ambitious widening. The original locks were only 110 feet wide; the plan called for them to be expanded to 180 feet wide. The plan also called for a dramatic increase in the depth of the canal, from 42 to 60 feet. Almost immediately, alarm bells went off in New York and New Jersey. It became clear that much of the container traffic from East Asia to the United States could be shifted to East Coast ports instead of landing on the West Coast and finishing the delivery by rail or truck.[4] With the widening of the Panama Canal, the pressure was on to raise the Bayonne Bridge. Design work began in 2011, and construction was started on the project, which was known as the Bayonne Bridge Navigational Clearance Project. The project was a joint effort of the Port Authority working with private contractors on something that had never been done before. They were planning on raising the height of the existing bridge without closing to traffic. Ships carrying $200 billion worth of goods were passing underneath the bridge every year, and there were 3.5 million vehicles crossing the bridge without any suitable alternative routes. Engineers could not shut down either the bridge or the shipping channel during construction. It was tricky, but engineers managed to keep the bridge open to traffic during peak hours. Lane closures were kept to weekends and nights. Meanwhile, the shipping channel remained open throughout the construction.[5]

Deepening the channel and raising the bridge all worked out. On September 7, 2017, the first Panamax ship, the *T. Roosevelt*, passed underneath the new Bayonne Bridge and entered the Port Newark–Elizabeth. "It was an amazing moment, not only because of the impressive size of the Panamax ship, but because this improvement will benefit our community," said Beth DeAngelo, a project director for one of the engineering firms. "One of the reasons that this is a great place to live is because of the accessibility to the large number of goods that enter through this port system. This ensures that these ports will continue to thrive as these larger ships become a standard mode of shipping."[6]

There were fears at first that the arrival of megaships would stress the terminals and lead to an increase of congestion. To be sure, cargo volumes did increase in 2018 as the vessel sizes grew, but predictions of severe disruption and delay did not materialize. "The real issue is we have not seen any major impact," said Jeff Bader, then-president of the Association of Bi-State Motor Carriers. He went on to explain that the volume of cargo had increased, but that the larger ships were not being loaded to capacity.[7]

One of the reasons that things went smoothly was that the terminals at Port Newark–Elizabeth were proactive in raising their cranes to accommodate the increased height of the containers on the bigger ships. In the long term, the U.S. Army Corps of Engineers will be looking at ways to improve navigation for these megaships as they move through the Kill van Kull. They will consider how to soften the bends in the channel to make it easier for the big ships to turn. They also want to figure out how to allow more than one large vessel at a time in the channel. "One way we could lift that restriction is [to] create a meeting or passing zone in the channel," said Bethann Rooney, deputy director of the Port Department of the Port Authority of New York and New Jersey. "There are certain areas of the channel that we could widen it through extra dredging to pass each other."[8]

Port Governance

A port authority in the United States is very different from a port authority elsewhere in the world. In Europe, for example, the cargo handling equipment, the line handlers, the tugboats, and the pilots are all part of the port authority. There, a port authority has the *authority* over all operations. On the other hand, here a port authority is only about handling the land and how that land is operated. Here in the United States, the pilots, the tugboats, and the line handlers are all private.

On October 8, 2019, I sat down with Bethann Rooney, the deputy director of the Port Department, in her office at Port Newark. She patiently explained the way things work: "As a landlord, we are renting the land. The terminal operators are responsible for the equipment that is operated on that land."[9] She went on to explain, as the recorder was running, that all of the cargo handling equipment including the ship-to-shore cranes, the trucks, the straddle carriers, the reef stackers, and the empty handlers are all privately owned by each terminal operator. Their employees do the driving and the handling of that equipment. In Rotterdam or Antwerp or Hamburg, by way of contrast, the people working all that equipment are public employees.

I found Bethann's explanation fascinating because the landlord model that she described preserves the American way of business. All these separate, private entities behave like profit-making corporations. For better or worse, Americans embrace private corporate ownership and free-market capitalism. Upward mobility, hard work, striving to get ahead—all of these are engrained in the American way of thinking. Our whole economic system is based on investments determined by private decisions and competition in a free market.

Of course, having this diverse mix of private ownership, with some public oversight, makes our story more complex. Bill Mongelluzzo of the *Journal of Commerce* explained that landlord ports "do not operate marine terminals, but they are neutral brokers that bring together stakeholders such as ocean carriers, shippers, terminal operators, truckers, railroads, and equipment providers."[10]

Measuring Productivity

With container ships getting bigger and bigger, ports worldwide including Port Newark are under tremendous pressure to improve their efficiency. Ports everywhere are being hit with huge surges in workloads. Since 2017, there has been a 16 percent increase in vessel sizes and a 19.4 percent increase in the average call sizes (containers being unloaded and loaded) during each vessel call.[11]

As the landlord, the Port Authority of New York and New Jersey wants to ensure that the land is being used in the most effective manner possible. The goal is to increase the amount of cargo that flows through their terminals. Here the Port Authority must look at different parts of the supply chain. First, how fast is the cargo being discharged from the ship? Second, how long does the cargo sit on the dock? Third, when a trucker arrives at the gate, how long does it take to get through the gate into the terminal, retrieve the cargo, and get that truck back out the gate and onto the street?

Measuring productivity gets complicated rather quickly because you realize that you must compare like things. For example, you can compare the productivity of one container terminal with another. You can compare the productivity of the three automobile processors—FAPS versus Toyota versus BMW. You can compare the productivity of bulk importers: salt versus scrap iron versus cement. However, you cannot compare the productivity of cement versus containers, nor can you compare cement versus automobiles.[12]

The Most Dangerous Two Miles in America

The attacks against the United States that took place on Tuesday, September 11, 2001, resulted in nearly 3,000 victim fatalities, over 25,000 injuries,

FIGURE 25 Green road signs in Jersey City, New Jersey, indicating proximity of Port Newark to the New Jersey Turnpike. Photo taken on April 8, 2018. The concentration of critical infrastructure here has led to the area being named "the most dangerous two miles in America." (Photo credit: Andriy Blokhin / Alamy Stock Photo.)

and the loss of some $10 billion in infrastructure and property damage. As a result, government attention turned to particularly vulnerable industrial areas around New Jersey Turnpike Exit 13-A, where there is a dense concentration of airplanes, railroads, trucks, and ships.[13]

Officials realized that Port Newark was among the most serious potential terrorist targets. In June 2002, less than one year after the attacks, the Coast Guard commandant joined President George W. Bush on an inspection trip of Port Newark. "Port security has always been part of the Coast Guard mission," Admiral Collins said. "But you wouldn't have called it a front-burner one. Now, of course, it clearly is."[14]

In November 2002, Congress passed the Homeland Security Act, creating the Department of Homeland Security as a cabinet-level department. That agency opened its doors on March 1, 2003.[15] Then, in 2006, the State of New Jersey organized its own Office of Homeland Security and Preparedness.[16] It was about this time that all of the security experts started referring to the area surrounding New Jersey Turnpike Exit 13-A as the most dangerous two miles in America.

FIGURE 26 A petroleum terminal and storage facility sits next to a railway and the New Jersey Turnpike in Linden, New Jersey, May 31, 2005, in an area named "the most dangerous two miles in America." (Photo Credit: Chip East / Reuters / Alamy Stock Photo.)

"It's the consequence that frankly scares the pants off of us, when you think about what might happen in such a congested area," said New Jersey Homeland Security director Richard Canas. He went on to explain an attack on one of the many chlorine plants in northern New Jersey could harm more than twelve million people within a fourteen-mile radius. He was also very worried about the port. He said that there were more than four million containers arriving there every year, but they are inspected only on the way out, not on the way in.[17]

In short, there is plenty to worry about in this area beyond just the port. Officials worry about the oil storage tanks and refineries, the pipelines, the air traffic, and the key highways. Despite all these worries, New Jersey has been spared any additional major terror attacks since 2001. I asked Bethann Rooney about this. This is how she replied as the recorder ran: "I take the term 'the most dangerous two miles in America' as a badge of pride. Ever since this area was labelled as such by the FBI, we have been able to prevent anything from happening. That is a source of pride for the Port Authority. In the immediate aftermath of 9/11, the Port Authority created a robust security organization. We have a very active program of awareness, prevention, protection, response, and recovery."[18]

Environmental Concerns at Sea

Every day thousands of shipping containers arrive at Port Newark full of consumer goods including T-shirts, computers, shoes, designer handbags, and toys. You name it; it probably came to you in a container. What is less obvious is that inside each container ship is another cargo—the heavy fuel oil that powers the engine and ends up in the air as exhaust.[19] For many years, the most commonly used fuel was "Bunker C," the lowest grade of fuel with such high viscosity that it required heating in order to pump it from the tank to the engine. Typically, this fuel had to be heated to 50 degrees centigrade or 122 degrees Fahrenheit before it could be pumped.[20]

Christian Eyde Moller, boss of the DK shipping company in Rotterdam, described bunker fuel as "just waste oil, basically what is left over after all the cleaner fuels have been extracted from crude oil. It's tar, the same as asphalt. It's the cheapest and dirtiest fuel in the world."[21]

Container ships that burned this fuel contributed greatly to air pollution. This bunker fuel was said to be the dregs of the refining process. It was black and thick as molasses, also loaded with sulfur.[22] Burning bunker fuel produces sulfur dioxide. When the sulfur dioxide combines with water and air, it produces sulfuric acid, the main component of acid rain. Environmentalists tell us that acid rain is a major cause of deforestation, and it tends to acidify waterways, harming aquatic life. Not only is sulfur dioxide bad for the environment, but it also has adverse health effects. It harms the respiratory function and increases the risk of infection. It aggravates conditions such as asthma and chronic bronchitis.[23]

Award-winning science writer Fred Pearce came up with a startling statistic that dramatized the problem of superships pumping out killer chemicals linked to thousands of deaths. He pointed out that just sixteen of the world's largest ships could produce as much sulfur pollution as all the cars in the world. Of course, the reason behind this startling statistic is that the ships could burn the cheapest and dirtiest fuels that cars on land are not allowed to use. Pearce figured that each of the largest ships could emit as much as 5,000 tons of sulfur in a year. That would be the same as 50 million cars, each emitting an average of 100 grams of sulfur in a year. With some 800 million cars driving around the world, that meant that 16 superships would emit as much sulfur as all the cars in the world.[24]

Despite all this grim news, there has recently been a major change. Starting on January 1, 2020, the International Maritime Organization (IMO) required that all fuels used in ships contain no more than 0.5 percent sulfur. This new cap represents a significant reduction from the previous sulfur limit of 3.5 percent. Public health experts estimate that this new cap will prevent about 150,000 premature deaths and 7.6 million childhood asthma cases around the world. "There are very few examples of air quality regulations that have as broad a reach of benefits as this one," said James Corbett, a professor of marine science and policy at the University of Delaware. "This is going to benefit children and adults in coastal communities located along major shipping lanes—not only [near ports] where ships are delivering cargo."[25]

The transition to the new rule went rather smoothly, since the IMO first announced the new rule back in 2016, giving shipping lines three years to prepare for the change. The carriers quickly realized that the low-sulfur fuel charges would have to be passed along to the customers. The costs of using low-sulfur fuel oil (LSFO) were real. There was an increase of about $150 for each 40-foot container shipped from Asia to the West Coast and a cost of about $250 for each 40-foot container shipped to the East Coast. Experts estimated that the new regulation costs the container shipping industry somewhere between $10 billion and $15 billion. Even so, there was very little pushback from consumers. "This is a non-negotiable surcharge," said David Bennett, president of the Americas at Globe Express Services. He went on to compare the new shipping charges to a common experience for everyday motorists. "They don't pull into the local gas station and fill up the tank with premium gasoline expecting to pay the price for regular gas," Bennett said.[26]

Environmental Concerns at the Port

The port serves the entire Northeast of the United States, but its immediate neighbors including Newark and Elizabeth bear an unfair share of the noise, pollution, and congestion that the port generates. To be sure, the neighbors have complained, and the Port Authority has listened. I asked Bethann Rooney about the problem. "We are an industrial complex," she said. "We are in close proximity to a number of densely populated residential communities. We need to be a good neighbor." Thus, environmental protection has become a top priority for the Port Authority. Back in 2009, the agency created a clean air strategy (CAS) for its port department. The CAS set out a goal to ensure a reduction in certain air pollutants such as nitrogen oxide, particulate matter, and sulfur dioxide while still growing the cargo volume.[27]

The Port Authority has been reasonably successful in doing this on a voluntary basis. Sulfur dioxide has been reduced by 98 percent. Particulate matter has been reduced by nearly 70 percent. In addition, greenhouse gas emissions have been reduced by 8 percent. All the while, the cargo volume has grown by 25 percent. The agency accomplished all this through a series of programs focused on the five sources of the pollutants: drayage trucks, locomotives, cargo handling equipment, commercial vessels, and harbor craft including tugboats and ferries.

With regard to the drayage trucks, the Port Authority could have simply forbidden polluting trucks to enter the port, but they took a different approach—one that was more carrot than stick. The agency routinely provides grants of $25,000 to licensed motor carriers to encourage them to eliminate trucks that are dirtier and more polluting, based on the age of the truck. The only requirements that come with the grant are that such a truck serves the port with at least 150 truck moves in a year and that the owner destroys the old truck.

When it comes to commercial vessels, the Port Authority encourages the owners to use low-sulfur fuels and to reduce their speed as they are entering and exiting the harbor. The agency provides cash incentives to ocean carriers that agree to these terms. Similarly, when it comes to cargo handling equipment, the agency provides incentives of cash grants to the owners to replace outdated equipment with clean diesel or electric equipment. The agency does basically the same thing with the locomotives, the ferries, and the tugboats. Of course, all of these programs are expensive, but they are important to the Port Authority.[28]

Given all the expense connected with these environmental programs, I asked Bethann Rooney if she would favor less stringent environmental regulations. This is how she put it as the recorder ran: "I favor more stringent regulations for the entire region." Rooney explained that, at present, the Port Authority is doing its entire environmental protection program voluntarily. Of course, the neighboring communities would like the Port Authority to do more.[29]

Unfortunately, even if the Port Authority were to spend $40 million to replace every truck doing business in the port, it would not make a difference for the health of the residents. The reason is that there are thousands of trucks going up and down the New Jersey Turnpike and into their communities every day that have nothing to do with Port Authority facilities. "We are living and working here just as much as our neighbors in these communities," Rooney said. "We are not going to move the needle on the health effect without more stringent environmental regulations."[30]

FIGURE 27 Trucks line up at the Port of Long Beach in California on October 15, 2014. A shortage of transportation equipment delays shipping containers and threatens the supply chain. At first, this was seen as a West Coast problem, but increasingly East Coast ports including Ports Newark and Elizabeth are dealing with the problem of congestion. (Photo credit: Lucy Nicholson / Reuters / Alamy Stock Photo.)

Congestion and Bottlenecks

The customers of today want shorter and shorter delivery times. That requires faster and faster supply chains. Meanwhile, the ships, terminals, trucks, and trains encounter problems with the result that the cargo moves slowly and unreliably. Research shows that container ships are on time only 75 to 80 percent of the time. When these ships are late, the average delay is three and a half days. If everything goes smoothly, it should only take two days from the time that the vessel arrives in port for the containers to exit the gate, but in many cases that might stretch to four or five days. All this takes place in an environment where customers now expect that the longest that they have to wait for a purchase is two days. Even speedier, Amazon sometimes advertises deliveries within two hours.[31]

Bethann Rooney wisely explained that congestion is a symptom, not a cause. Thus, the Port Authority has looked carefully at the causes of congestion. Often the cause of an apparent piece of congestion is a bottleneck somewhere else in the system. It is important to look at the entire supply chain, and ocean carriers are at the beginning of the supply chain. The ability of a given ship to reach the port on time, when it is scheduled to be there, has an impact on what happens at the terminal, with regard to the railway and with regard to the truckers. There is a cascading effect if any part of the supply chain in not operating efficiently. We can think in terms of the gears of the supply chain. All of these gears have to working in precision. Rooney explained, "If there is one bit of sand that messes it up, the whole system shuts down."[32]

Starting in 2014, the Port Authority under the leadership of Bethann Rooney established the Council on Port Performance (CPP). This council is a group made up of both the Port Authority and the New York Shipping Association. The idea was to bring the industry leaders together to help tease apart the supply chain and understand what was going on. In the past, terminal operators were concerned only with what happened within their own fence lines. They did not fully understand the implications of decisions that they made on truckers. Nor did the ocean carriers understand or care that, if they were off schedule, there would be cascading effects upstream and downstream.

"We are focused on the issues," Rooney said. "And we have everyone rowing in the same direction. The CPP is not the ultimate authority, but at least the discussion is happening. There is transparency and communication. The silos are being broken down."[33] In other words, there is an understanding that if the port is inefficient, it hurts the terminal, the ocean carrier, the trucking company, the warehouses, and even the chassis provider. Everyone is going to lose money if the port is inefficient.

On a positive note, there has recently been a tremendous amount of investment that the terminal operators have made in new equipment, software, and infrastructure. They invested over $3 billion in the ten-year period running from 2009 to 2019, with another $2 billion projected to be spent from 2019 to 2029. The terminal operators need the right systems, processes, and infrastructure in order to keep the supply chain productive. Their profits are at risk if they do not make the necessary investments.[34]

Minimizing Theft

In the era before the 1960s, as we have seen, cargo was moved according to break-bulk principles. In other words, cargo was packaged in cases or on pallets. It was loaded and unloaded on a piece-by-piece basis. It was easy for cargo handlers to steal electronics, appliances, clothing, engine parts, cosmetics, liquor, and cigarettes during loading and unloading. When containers were first introduced, theft was greatly reduced. However, as Barry Tarnef of Chubb Insurance explained, "After an initial honeymoon period, during which criminals adjusted to the new container system, other patterns of theft developed."[35]

At the present time, in a given year, the international shipping industry transports some 130 million containers packed with cargo. Most of the time, these shipments reach their destinations without any trouble. Still, cargo loss does take place. Here, we will look at three possible sources of loss—sea transport, port handling, and finally truck stops and warehouses.[36]

Moving cargo by sea is relatively safe in terms of theft, compared with the risk ashore. To be sure, piracy is still an active threat, particularly in Southeast Asia, but that threat has significantly decreased in recent years. Why is shipping at sea relatively safe? Security expert Jim Yarbrough said, "We do not typically see many incidents of theft involving maritime cargo in transit. There are much fewer vulnerability points on transport once it is at sea. The real risk is at the point of loading or off-loading, where we do see greater numbers of thefts and attempts."[37] Containerized cargo, loaded onto a large vessel, moves along at 24 knots, in isolation and over long distances, far away from shore. All of this makes for reasonably secure voyage in terms of theft avoidance.

Once the cargo arrives at one of the terminals at Port Newark, it is mostly secure. There is rarely the loss of an *entire container* from a terminal because of the security measures that are in place. However, there is always the potential for pilferage, the loss of some of the contents of a container. When a cargo container is originally loaded, there is a certificate issued of what went into the box. The box is then sealed, and there is an expectation that, when the box gets to its final destination, the seal will still be intact. However, there are instances when a door is taken off or a vent is taken off. Things can be put in the box that are not supposed to be there, and things can be removed that are supposed to be there.[38]

In my research, I have found no reported instances of theft from Port Newark. I could not find any examples either in print or online. Granted, Port Newark is very safe, but why would there be no reports of theft at all? With persistence, I found a government study that helps to explain the absence of any reports of thefts at terminals throughout the United States. In other words, Port Newark is not unique in reporting no thefts from containers. It seems that shippers are reluctant to report a theft or a loss for a number of reasons.

Primarily, reporting a loss would have a detrimental effect on the reputation of both the shipper and the port. Besides, most shippers are self-insured for an amount of up to $500,000, so small losses can be chalked up to just the cost of doing business. Equally important is the fact that filing a claim could easily result in an increase in the shipper's premium. In addition, experience has shown that reporting a theft to the police and then following up in the courts is seldom satisfactory. Shippers have found that following the correct legal procedure rarely leads to a substantial outcome because the courts tend to see cargo theft as victimless crime. Even when convicted, the thieves often receive minimum fines and minimum incarceration, sometimes even dismissal.[39]

Insurance experts agree that cargo is relatively safe both at sea and upon arrival at port terminals. The real problem begins when commercial freight trucks leave the port terminals. New Jersey is among the top states for theft because it has a very large port, rail depots, and interstate highways. New Jersey also has a large population density among which stolen goods can be readily distributed.

Thieves tend to target truck stops for obvious reasons. When drivers stop en route for a cup of coffee or a trip to the restroom, they are leaving the truck unattended. Of course, they should do a walk-around inspection of the tractor and the trailer when they return to their truck, but this is seldom done. Thus, they will not realize that something is missing until they arrive at their destination. By this time, the cargo is likely to have been sold on the black market.

Warehouse stops present similar problems. Thieves may stake out the site for weeks and even months. They may use insiders to tell them about loads and delivery times. Anytime that the truck is unattended, it is vulnerable to theft. Thieves are looking for household goods and electronics that can easily be sold. Increasingly, warehouse operators are installing stronger security fences and surveillance cameras as well as hiring additional security guards.[40]

Food and beverage products are often stolen because they are in demand on the black market. Theft specialist Scott Cornell said, "It is easy to sell because everyone needs to eat. It's also less traceable and easy to hide because once it's consumed, it's gone."[41]

Contraband in Containers

On any given day, there are at least three federal agencies at Port Newark looking for contraband in containers. First, there is the Consumer Products

Safety Commission (CPSC), which is looking for unsafe goods. This agency is charged with protecting the public from the risk of injury or death from many products. Second, there is Customs and Border Protection (CBP), which is looking for fake name-brand goods. Third, there is the Drug Enforcement Agency (DEA), which is looking for illegal drugs, especially cocaine. Let us take a brief look at the work of each.

The CPSC is especially busy in the period leading up to the Christmas holiday. They are on the lookout for things like toy cars that are made of lead and Christmas lights that are prone to catch on fire. They are also looking for toys that are coated in lead paint and play sets that might pose a choking hazard. These hazardous goods often show up at discount stores or as sales on the street. The CPSC is the first to admit that they do not have the funding or the staffing to catch this flood of unsafe goods. They said, "What's keeping us from doing that is funds."[42]

Apart from dangerous goods, other officials are on the lookout for counterfeit merchandise. CBP addresses this problem. One the one hand, the agency is constantly looking for counterfeit brand-name children's merchandise such as Betty Boop dolls, Pokémon trading cards, Dora the Explorer watches, and Winnie the Pooh bubble blowers. On the other hand, the agency is looking for counterfeit adult merchandise such as Gucci bags, Lacoste shirts, Rolex watches, and Hermès bags.[43]

A recent well-publicized case was the seizure of $1.7 million in fake Nike sneakers at Port Newark in 2018. "This significant seizure of counterfeit Nike sneakers illustrates Customs and Border Protection's continued commitment to protecting the American consumer against the proliferation of substandard and potentially unsafe counterfeit consumer goods," said Troy Miller, director of CBP's New York Field Office. "Enforcing Intellectual Property Rights laws is a CBP priority trade mission. We will continue to work closely with our trade and enforcement partners to identify and seize counterfeit merchandise that could possibly harm U.S. consumers and businesses."[44]

Meanwhile, there are millions of containers that are moved around the globe by sea every year. Most of these are carrying legitimate cargo, but some are used to smuggle drugs.[45] According to the DEA, smugglers are increasingly using legitimate commercial freight containers to conceal cocaine and heroin. They have also been found to use bribery or extortion to corrupt port workers, police, and customs officials. Over the past fifty years, the United States has engaged thirty-seven federal agencies in the

FIGURE 28 U.S. Customs and Border Protection conducts vessel and cargo container inspection at Port Newark on November 6, 2012. The agency is one of the world's largest law enforcement organizations, and it is charged with keeping illegal goods out of the United States, while facilitating lawful international travel and trade. (Photo credit: Alamy Stock Photo.)

war on drugs, including the DEA, CBP, the FBI, the Coast Guard, and the Secret Service.[46]

These agencies have had some success, but the U.S. State Department reported, "Worldwide narcotics production is reaching new levels, corruption continues to undermine enforcement efforts, and a number of governments still fail to exhibit a serious commitment to reducing drug production and trafficking."[47]

Port Newark handles thousands of ships and millions of containers every year, but customs officials are able to inspect only about one percent of all container traffic. The principal drug seized at Port Newark is cocaine. It has been found in an amazing number of places including plastic tubing, diesel engines, walls and ceilings of the containers themselves, fruit juices, walls of cardboard boxes, flowers, chocolates, and even plastic bags inside the stomachs of live fish. Smugglers rely on the high volume of container traffic at Port Newark to provide a statistically low probability of detection. In addition, backlogged containers that sit idle give thieves additional opportunities for surreptitious pilferage.[48]

FIGURE 29 Container cranes of the Port Newark Container Terminal (PNCT) located on the north side of the Elizabeth Channel on July 27, 2011. The terminal occupies 272 acres, and it handles more than 1.49 million TEUs annually, as this is written. PNCT secured a long-term lease agreement with the Port Authority of New York and New Jersey through 2050. (Photo credit: Clarence Holmes Photography / Alamy Stock Photo.)

Size as a Limiting Factor

At the port, there are three major container terminals. Ranked by size, we find Maher Terminal the largest, with 450 acres. Next, is the APM Terminal, with 350 acres, and third is the Port Newark Container Terminal (PNCT), with 272 acres.[49] Although this adds up to a considerable amount of acreage, a glance at the map shows that there is almost no room for growth. The port is constrained by Newark Bay to the east and by the dense urbanization and the New Jersey Turnpike to the west.

I asked Bethann Rooney if she sees size as a limiting factor for the growth of the port in the future, and she replied that it is not. "We can use the property more efficiently in order to get more capacity out of it," Rooney said. She went on to explain that this efficiency can be accomplished by using more automated technology, by stacking the cargo higher, and by using an appointment system for the trucks. Equally important is the possibility of adding hours of service. As this is written, the terminals in Newark and

Elizabeth are operating only from six in the morning to four, five, or six in the evening, during weekdays. Rooney explained that there is much more time in the day and in the week to move cargo through the terminals. Of course, expanding the hours will depend greatly on negotiations with the International Longshoremen's Association (ILA).[50]

Another possibility for overcoming the size limitation would be to eliminate bulk cargo service from Port Newark–Elizabeth and move that operation elsewhere, opening up more room for handling containers. In fact, there is a facility in Carteret called Port Reading that has been sitting idle for a long time. Potentially the bulk cargoes could be moved there, since bulk cargoes do not need the deep water of some 50 feet that we find at Port Newark. Port Reading has a depth of some 35 or 40 feet, quite adequate for handling bulk cargoes. Moving the bulk cargoes south like that would create more land at Port Newark for containers.[51]

Superstorm Sandy

Hurricane Sandy started as a tropical disturbance off the coast of West Africa on October 11, 2012. It entered the eastern Caribbean Sea on October 18, and by October 24 it officially became a hurricane 90 miles south of Kingston, Jamaica. The hurricane went north along the southeastern United States, and it intensified. It made landfall on October 29, and it drove damaging storm surge and waves into New York Bay and up the Hudson River, causing urban flooding in Jersey City and Hoboken, New Jersey.[52] The storm had a tremendous impact not just on Port Newark and Port Elizabeth, but also on the entire Port of New York and New Jersey, which was closed by the U.S. Coast Guard on October 28. The port was not reopened to vessel traffic until November 4. Even then, although the waterways were open, most terminals were unable to resume operations for weeks because of power failures and equipment damage due to the storm surge, plus a high tide.[53]

The closure of the entire port for cleanup cost the economy billions of dollars. Terminal infrastructures were damaged, and thousands of new cars were flooded. Yard equipment was ruined, and debris on roadways had to be removed. There was loss of power for equipment and road signals. While the port was closed, the containers that were supposed to be sent to Port Newark were diverted to Halifax, Canada, and to Norfolk, Virginia. These containers were then trucked to their final destination.[54]

I remember Superstorm Sandy very vividly. Compared with those whose homes were destroyed, my setbacks were minimal, yet hardly forgettable. My house in New Jersey was without power for ten days, and the storm destroyed the fence surrounding my back yard. It set back my research at Port Newark for weeks because the port was closed to visitors while the damage was being repaired. What can be done to return the port to its normal operation after such a massive disruption? I asked Bethann Rooney about lessons learned, about mitigating the loss from such hurricanes in the future.

"These things are on our radar screen, but these things are difficult to manage," Rooney said. "You cannot build yourself out of the problem. When we look at the predictions for a one hundred year storm, with its accompanying sea-level rise, we would have to raise the height of all of our docks somewhere between eight to twelve feet."[55]

Rooney went on to explain that, in an operating port, you could not just raise the docks that way. The work would also require raising all the land behind the docks. Such a maneuver would be completely disruptive, not to mention that in the end you would have a port facility that would be too high for the local roads and infrastructure. Of course, you might be able to prepare for a catastrophic sea-level rise in a new facility, building it from scratch. Having said that, there are piecemeal operational measures in the post-Sandy world that all the terminals have undertaken.

Driving around the port today, you can see that the terminals have elevated their generators and substations beyond flood level. The control boards for the traffic lights were all infiltrated with water during Hurricane Sandy because they were all at a three- or four-foot level, for worker convenience. Now the control boards are at the eight- or ten-foot level. It is safer, of course, but it presents a more difficult maintenance responsibility because now the terminals must have a man lift at each control board. At least the control boards are now out of the flood zone.

Years ago, all of the ship-to-shore cranes were diesel powered. In the modern, environmentally aware period, these cranes were converted to electric power to reduce pollution. However, there was an unforeseen problem. The new electric cranes do not do well with water. In Hurricane Sandy, they were all inundated with storm water. At great expense, every engine had to be taken out, cleaned with fresh water, baked and dried, lubricated, and then put back into the crane. The port was a victim of its own environmental success when it came to a natural disaster.

Rooney summed up the Port Authority's plan for dealing with future hurricanes: "With existing infrastructure, we can take small operational measures such as raising things out of the flood zone. We can also take proactive steps when we learn that a storm is heading our way. We can take the engines out of the cranes, and we can take the motherboards out of the traffic control systems. This is how we plan to survive future hurricanes."[56]

To be sure, the Port Authority is doing all that it can to prepare for these catastrophic events, but experts warn us not to become complacent. "Hurricane Sandy was obviously a terrible event for the Northeast United States, but it really was not the worst-case scenario," said Dr. Megan Linkin, a natural hazards expert. "The East Coast is not immune to a hurricane that brings a Sandy-like surge and extreme winds over a large area."[57]

Increased Awareness

In normal times, most of us give little thought to our reliance on imports delivered with container shipping from overseas. But, once in a while, something comes along that serves as a reminder. Such was the stay-at-home period due to the ongoing COVID-19 pandemic. There were government regulations that varied from state to state, regulating things like mask wearing and social distances; but these regulations were seldom enforced with strictness. Even in a pandemic, some people, deemed essential workers, had to go work. These included health care workers and grocery store workers. By way of contract, many people simply stayed home voluntarily out of an abundance of caution. It was made easier for them if their employers insisted that they work from home.

At first, the stay-at-home people felt a sense of relief. They were spared the need to dress up and commute to work. They could work in comfortable clothes at their computers. However, over time, they began to feel stressed out because they could no longer visit in-person with friends, patronize bars and restaurants, or attend movies or sporting events. They were dining at home for every meal; they were watching more and more television; the children said that they bored. They began to experience what has been called "pandemic fatigue" or "pandemic burnout."

What to do? All at once, many turned to online shopping for consumer goods. Suppliers all over the world received a surge of new orders. Container ship operators suddenly had more business than they could comfortably handle. Port Newark and Port Elizabeth had to deal with increased congestion and a shortage of shipping containers. People working at home were converting underused spaces into home offices, ordering desks and chairs and computers for their workspace. They were ordering home gym equipment like elliptical machines. They were ordering expensive French wines by the case. For their kitchens, they were ordering baking supplies and expresso machines. This surge in imports did not go unnoticed in the press. A *New York Times* story of March 2021 quoted Bethann Rooney as saying, "Never before have we had anything like that. The cargo was coming fast and furious into the country."[58]

Imports were no longer just being taken for granted; there was an increased awareness of our dependence on them. A good example of this phenomenon is the case of Hattie Kolp, a special education teacher at a Harlem school. As she began teaching remotely, she turned her guest bedroom into her office, with a green velvet couch made in Asia that she had ordered online. She began eagerly tracking her couch as it was carried across the Pacific in a container ship, then delivered at a West Coast port before being transported by rail to Manhattan. She said, "I think it just shows us that very little is made in the U.S. It's really bringing attention to how far stuff travels to get to us."[59]

10

Moving the Freight

• • • • • • • • • • • • •

On the Jersey side of the larger Port of New York and New Jersey, there are five container terminals. Ranked by size, we start with the Maher Terminal in Port Elizabeth at 450 acres, followed by the APM Terminal in Port Elizabeth at 350 acres. Next is the Port Newark Container Terminal (PNCT) with 272 acres, followed by the Global Container Terminal with 167 acres over at Bayonne. The smallest is the Red Hook Container Terminal at 30 acres just north of PNCT at Port Newark. This last one often confuses people. You might say, "Wait a minute. Red Hook is in northwest Brooklyn." Well, you would be correct. The main Red Hook Terminal with some 80 acres is indeed in Brooklyn. However, Red Hook Terminal is the only operator in the Port of New York and New Jersey with facilities in both states. The smaller of the two is at Port Newark, where they operate a cross-harbor barge service that goes back and forth across the Hudson River daily. Over the years, I have had the opportunity to visit all five at least once.

For a ship buff like myself, such visits are thrilling and exciting. At the same time, these visits can be confusing. There is so much going on, the experience can be overwhelming. We find giant ships and loud trucks and beeping work vehicles and workers with hard hats and reflective vests everywhere. There are stacks of big boxes in many colors. We see giant gantry cranes looming overhead. There are audible alarm sounds going off. How to make sense of all this confusion? There are many things happening simultaneously, so it makes it hard to tell the story in a coherent fashion.

I have found that to make sense of it all, it is important to remind myself that there are really just five basic areas. First, there is the gate complex on the land side where the motor carriers enter and leave. Second, there is the water side where we find the berths to which the ships are secured to the dock. Third, there is the immediate landside area, or pier, where we find the cranes that load and unload the ships. Fourth, there is the container yard where containers discharged from a ship are stored and where containers to be loaded onto a ship are stored. Fifth, there is the on-dock railroad yard.[1]

With this five-area model in mind, I began to see that all five of the container terminals on the Jersey side have many things in common. They all can be considered as intermodal interfaces in the global transportation network. Basically, they serve to connect container vessels on the sea with trucks and railroads on the land. They all need to find ways for efficient handling of containers, and they need to deal with the storage space allocation problem. All five need big cranes to get the containers on and off the ships. They all need to have some kind of horizontal transport to get the containers to and from the storage stacks. They all need to be able to somehow stack the stored containers so as to save on space.[2]

All of the terminals on the Jersey side are dealing with the challenge of handling mega-vessels carrying more and more containers. Shipping companies are looking for reliability to stick to delivery dates. All of these terminals must deal with the problem of distributing or reusing empty containers. All must grapple with the cost of replacing older equipment with more efficient ones because they are under pressure to handle more containers in a shorter time at a lower cost. And then there is the manpower planning problem. How many workers do you need at a given time to ensure continuous efficiency of the terminal? To address these problems, all must increasingly rely on new software to manage port operations. Increasingly, terminal managers are turning to operations research to aid in problem solving and decision making.[3] Although these terminals have much in common, each one has its own individual history and its own way of doing things. Each one has an interesting story; however, for our purposes, let us focus on the one that has evolved from the heritage of Malcom McLean.

APM Terminal at Port Elizabeth

Earlier in this book, I told the story of how Malcom McLean invented containerization at Port Newark way back in 1956. I then went on to explain the

FIGURE 30 Container cranes at the APM facility with 350 acres at the Elizabeth Marine Terminal, also known as Port Elizabeth, July 27, 2011. Since this photo was taken, the cranes have been upgraded. They now have four ultra-super-post-Panamax cranes. These cranes can reach the twenty-five containers across that can be found on the ultra-large container ships. (Photo credit: Clarence Holmes Photography / Alamy Stock Photo.)

subsequent construction of the Elizabeth Marine Terminal in 1958. Then in 1962, Governor Richard J. Hughes announced that the Port Authority had signed a twenty-year lease of its new facilities at Elizabeth with McLean's Sea-Land Service. Here was a brand-new port, costing millions of dollars, that was entirely dedicated to handling containers. Over time, under McLean's management, the new port proved successful, notwithstanding the economic ups and downs. Finally, in 1977, McLean left the company, whose capital was sold to Maersk in 1999. The name of the company was changed to Maersk SeaLand, and it finally became the Maersk Line in 2006.[4]

I was eager to learn how things had evolved since then, so I made an appointment to visit Giovanni Antonuccio, general manager of client services for the APM Terminal in Port Elizabeth. Why the name APM? I was told that it stood for Arnold Peter Møller, a Danish businessman who founded the A.P. Moller-Maersk group back in 1904. Later I learned that APM Terminals is the terminal operating division of the Maersk Group, the parent company of the world's largest container shipping line by capacity, the Maersk Line. They have some seventy-two terminals in sixty-nine countries, all over the world.⁵ I received a warm welcome from Mr. Antonuccio at the Executive Offices on the third floor of the APM office building.

I felt fortunate to see this place, especially in light of the heightened security of our post-9/11 world. It's a place that few outsiders ever get to visit. As I looked around, I kept thinking: would Malcom McLean recognize the terminal? So much has changed over the years. I had many questions, and Giovanni filled me in patiently. It turned out that the biggest customer today at this terminal is the Maersk Line, including Safmarine and SeaLand. Thanks to something known as the Vessel Sharing Agreement, there are several other carriers that also stop here. I asked Giovanni, "What gets imported here?" He replied, "Well, since the United States is primarily a consumer nation, we pretty much import everything—from electronics to food stuffs, to clothing, and car parts. Our biggest export is the empty container. We also ship out waste paper and scrap metal."⁶

A common problem all over the world for seaport terminals is that, as the ships get bigger, the number of containers gets larger. However, it is difficult, if not impossible, to expand because seaport cities are already built up and the land around them is prohibitively expensive. APM Terminal in Elizabeth is no exception to this problem. I asked Giovanni how his company has been dealing with the issue. He explained that they cannot make any more property. All they can do is make better use of what they have, so they launched a $200 million upgrade with three components. First, they had to upgrade the technology, to find the proper operating software systems to handle cargo. They decided on a system called Navis N4, which is described as a technology platform that helps with the intelligent movement of goods. It is said to optimize container handling, vessel planning, and yard utilization.⁷

Second, they had to upgrade their cranes, so they now have four ultra-super-post-Panamax cranes. These cranes can reach twenty-five containers across and ten containers high on deck, which you find on the ultra-large container ships. Third, they launched a new gate project with twenty-four inbound lanes and sixteen outbound lanes, which have the latest technology. He went on to explain that new gates use RFID (radio frequency identification) and OCR (optical character recognition). In other words, the cameras will be reading the numbers off the container and the chassis, and the clerk will simply verify if there is a match and handle any exceptions. The new gates will be able to weigh the truck, even as it is moving. The driver will eventually be able to process with minimum voice interaction, all by use of appointment, RFID, and OCR, creating a faster transaction between the gate clerk and the truck driver.⁸

From my previous experience having talked with truck drivers who were in the business of dropping off and picking up containers at Port Elizabeth, I knew that their principal complaint was the amount of time wasted at the terminals, not just at APM, but at all of them. It seems that the problem stemmed from having too many trucks arrive at a busy terminal. For many years, the policy was simple: first come, first served. But it involved a good deal of waiting around for your turn. The obvious solution was to implement an appointment system, but it was easier said than done. Anytime you put a new system in place, there is danger of confusion and disruption. APM at Port Elizabeth launched their appointment system on February 29, 2020. The system was rolled out slowly and carefully, starting on Saturdays only because normally all of the terminals in New York and New Jersey are closed on Saturdays. If, for any reason, the system failed, there would not be a major disruption. Truckers could still come into the terminal without an appointment, except for Saturdays, at this point. By March 13, they shifted to just Fridays for the appointment system. They then added a day each week as long as there were no major issues. Fortunately, there were no major issues.[9]

Clearly, the goal of the APM Terminal is to optimize terminal operations by getting large numbers of containers to and from a given ship's side. However, there are many variables that enter into the equation, making that goal difficult to achieve. It helps in planning if the terminal knows in advance when a given ship will arrive. Of course, the terminal negotiates with a given shipping line to work out an agreed-upon schedule. They might agree, for example, that the shipping line will have its ships arrive every Tuesday afternoon. Thus, the terminal will reserve a berth for that ship at that time. If the ship arrives on *pro forma* (agreed schedule) as per the contract, the terminal must make a berth available or else they are responsible to compensate the line for delays through various measures. That is the plan, but things can go wrong. Perhaps the ship encounters bad weather or perhaps the ship develops engine problems. If the ship arrives late, it disrupts the plan. The terminal will do its best to accommodate the late ship, but it must first honor the agreements with other ships that have arrived on time.[10]

Let us take up the procedures for the arrival of a typical ship at the terminal. To keep things simple, let us assume that our ship arrives on time. In this example, let us assume that our ship is coming from Rotterdam to Port Elizabeth. For planning purpose, it would be nice to have a cargo manifest, listing all the containers onboard in advance. However, in this hypothetical case, our ship is first going to stop in Halifax, Nova Scotia. There some containers will be removed, and others will be loaded. Thus, we must wait until some twenty-four or forty-eight hours to get the updated list. Then we will get a BAPLIE, which stands for bayplan including empties. It's a message that tells the receiving terminal the exact stowage position and container number of the cargo onboard the vessel. Obviously, this is very valuable information for preplanning the discharge of the containers.[11]

Once the ship is safely docked, the terminal can begin simultaneously unloading and loading the ship by means of a number of specialized cranes, known as shoreside gantry cranes, or ship-to-shore (STS) container cranes.[12] These cranes, sometimes called wharf or dockside cranes, can lift or move heavy loads by means of projecting an arm, or boom, that reaches across the ship. The largest of these cranes can reach across twenty-five containers on a ship, piled ten high. Located on land immediately next to the water, these cranes can handle hundreds of containers a day. These cranes are called gantries because they are moveable. They move sideways, on thirty-two steel wheels, along specialized tracks built into the dock, allowing them to work different areas of the ship. Several cranes can work each vessel at the same time, optimizing the turnaround time. Attached to the boom that moves across the ship is a trolley that houses the operator cab. A given lift of a single container is supposed to take only two minutes. That adds up to thirty moves per hour if all goes as planned.[13]

The way it works at the APM Terminal in Port Elizabeth, for loading a ship, is that the crane operator can pick up one 40-foot container or two 20-foot containers at a time from a "bomb cart," which is a heavy-duty trailer with solid tires. The bomb cart is basically a modified flatbed trailer with twelve-inch walls, making it a sort of tub. The container simply sits in the tub; it is not bolted or attached. This arrangement saves time. There is no need for a chassis, because the bomb cart is neither licensed nor approved to travel over the road. They have the designation of "bomb carts" because they are "bomb-proof," or rugged, designed to withstand heavy containers being dropped onto them suddenly. The bomb carts are pulled by "hustlers," which are small trucks that stay inside the port. The yard hustler truck gets its name from the verb "to hustle" in the sense of "to hurry up" or "to work hard." These trucks are very different from the over-the-road trucks. They are designed for a single operator with a 360-degree view. They cannot travel at highway speeds, so they usually have a top speed of only 25 mph (due to a mechanical limitation placed in the units by the terminal). The loaded hustlers are constantly bringing containers from yard stacks to dockside to be loaded onto the ship. At the same time hustlers are bringing containers discharged from the ship to stacks in the storage yard. In other words, the hustlers are the mechanism for horizontal movement of containers to and from the ship. Typically, one might see one crane loading 40-foot containers *onto the ship*, as they are taken off the bomb carts one at a time. Meanwhile, there might be another crane unloading 40-foot containers *off the ship* onto bomb carts one by one.[14]

Using these huge STS cranes requires considerable teamwork. Typically, each crane has one container gang, made up of ten longshoremen. They are all members of the ILA or the International Longshoremen's Union, AFL-CIO, the largest union of maritime workers in North America.[15] At any one time, there is one foreman who serves as the boss, one crane operator, and a number of holdmen. The holdmen are dock workers who guide the hustler trucks into the best position for the crane to pick up the containers. They make sure that the crane properly locks onto the boxes for loading and unlocks properly for unloading. They also maintain communication with the crane operator by means of both radio and hand signals. The ten-man gang, of course, also has a group of hustler drivers that supports them during the stevedoring operations. These hustler drivers bring containers to and from the yard, to and from the crane. The holdmen are responsible for signaling the crane operator and lining up the hustlers under the crane, while the gang foreman oversees the operation to ensure that it is safe and productive. If there are any problems, the gang will notify the superintendent as well as head foreman designated for that particular vessel if it is something that cannot be solved quickly on the dock. This can occur when the vessel crew requests a change to the stowage plan, or if damage is noticed to either the vessel or the cargo being handled.[16]

Up to this point, we have been explaining how the terminal uses STS cranes to load and unload containers. The process goes smoothly because the hustler trucks are constantly bringing containers from the stack to the wharf at the water's edge to be uploaded onto the ship. Meanwhile, other hustler trucks are constantly taking the offloaded containers from the ship to the stack. We have been looking at the busy activity at the wharf. But meanwhile what has been going on at the yard? We can think of the yard as the temporary storage area where inbound and outbound containers are stored. The incoming containers arrive as one of two types; specifically, there are 20-foot containers and 40-foot containers. Thus, a 20-foot container occupies one 20-foot equivalent unit (TEU). And a 40-foot container occupies two TEUs.

For the APM Terminal in Port Elizabeth, the yard is said to have a storage capacity of 69,736 TEUs or some 34,868 40-foot equivalent units (FEUs).[17] In an ideal situation if the port had unlimited space, they could spread out the containers, one by one, over acres of space. If that were the case, it would be easy to pick them up, one at a time. But in the real world, space is limited, so you have to begin stacking the containers. The containers can then be stored in a number of pads, or groupings, or blocks. The pads are designated areas with reinforced surfaces to withstand the weight of stacked containers. The containers are arranged in bays, lengthwise or end to end, up to 60 containers long. They are also arranged stacks, side by side, usually six across and four high in a grid pattern. To access the containers in these stacks, we need a specialized kind of crane. This is a yard crane as opposed to the quay crane, or STS crane, mounted on rails, that we have already discussed.[18]

What we have in the yard is the rubber-tired gantry, or RTG. It is much smaller than the STS crane, and it is more mobile because it can travel on its rubber tires. This is a diesel-electric bridge crane used to move containers. It has a main frame made of two large girders, and it is supported by four leg columns. Each column rests on an array of four tires, for a total of sixteen tires. It supplies its own electrical power that comes from a diesel engine. There is an operator cab positioned high on the crane, with maximum visibility even below. The operator can use the spreader to grab onto a given container. The spreader engages, adjusts, and moves the containers by means of hoist cables. The RTG can lift two 20-foot containers, or a twin pick, with a weight of up to 65 tons. Or it can lift a single 40-foot container with a weight of up to 50 tons. It can hoist and move hundreds of containers every day. It can span across several container rows. It can move a container up and down as well as from side to side. It can load and unload containers to and from the hustler trucks, as well as to and from over-the-road trucks. In normal operation, the RTG is reliable and safe; however, the operator must be alert to sudden changes in the weather. High winds can make things difficult and may even lead to catastrophic damages. As the wind speed increases, more and more caution is needed. Once the wind speed hits 45 miles per hour, an alarm goes off and the operation has to be postponed.[19]

As we have seen, stacking boxes is the best way to save valuable yard space; but once you depend on stacking, things get complicated quickly. Every time a box is added to a stack, a careful record must be made of its location. At the same time, the stacks are dynamic, constantly changing. The beauty of the RTG is that it makes stacking possible. With the aid of powerful software, the RTG "knows" the location of every box. Thus the RTG can "find" the box needed to be sent to wharf for export; it also can "find" the box needed to be put on an over-the-road truck and sent out the gate.[20]

To make things run smoothly requires a great deal of planning. For example, each incoming container should be assigned a location in the stack that is available. At the same time, ideally each outgoing container should be easily accessible. Thus, it is important to plan that containers on top of the stack should be those scheduled to leave earlier. It is not good if the RTG must first remove unneeded containers to get the desired one way down at the bottom of the stack. This is called reshuffling, and it slows things down. A review of the technical literature shows that there are many variables to be taken under consideration. For example, it is important to know the expected departure time (EDT) of the container. When is it to be removed from the stack to leave the terminal, whether by vessel, train, or truck? If we know the EDT, we have a better idea of where to place the container in the stack. Let's take another example concerning an outgoing ship which handles exports. That ship is likely to have more than one port of destination (POD). To properly load the containers, we need to know the sequence of the PODs. Clearly, the container that will be discharged in the first POD will have to be loaded onto the ship last. The examples go on and on, and the problem gets more and more complicated. Thus, port managers must be on top of their game, dealing with mathematical formulas, artificial intelligence, and software.[21]

In recent years, there has been more and more pressure on terminal operators such as APM to improve productivity. An obvious way to do that is for the operators to turn to increased automation. At the same time, that trend has rattled the ILA. An interesting agreement was reached in 2018. There was a new system, agreed upon as part of a contract between the ILA and the New York Shipping Association (NYSA) on September 25, 2018. Under the new system, each container gang, under the leadership of the foreman, is expected to increase its productivity each year by one container move per hour, in order to reach a port-wide productivity of at least thirty crane moves per hour. Teams that do not meet the goal will be carefully examined as to what went wrong, and they will get help to fix their problems. In the worst possible case, an underperforming team might be given fewer hours.

Harold J. Daggett, the president of the ILA, who signed the agreement, sees it as labor's pledge to work more effectively and to hold off too much automation. At the same time, John Nardi, the president of the NYSA, has been pleased with the agreement because it holds the gang, and the foreman in particular, responsible for the team's productivity. He gave an example of how it is now possible to confront the foreman, "Okay, what is the problem? Is it the crane operator? Is it the people in the gang that are not doing what they are supposed to do? Is it the drivers not supporting the gang? Or is it an absentee problem." In an extreme case, a gang that is found to be consistently underperforming could be given fewer hours.[22]

Motor Carriers

For most of us, when we think of the job of truck drivers, we think of the popular culture image of long-haul drivers. They are the ones who transport large cargoes across state lines, often for weeks at a time. They do not have to worry about office politics. They get to see new cities and new states all across America.[23] Some of us remember how the romance and freedom of this lifestyle was reflected in the sudden popularity of truck driver songs as a subgenre of country music in the 1970s, with songs such as "East Bound and Down" by Jerry Reed. This song spent sixteen weeks on the *Billboard* charts back in 1977, topping off at number two.[24]

While spending time at Port Newark–Elizabeth, trying to figure out things work, I quickly learned that the popular culture image of the over-the-road truck driver did not fit what was going on. The trucking service that carries containers a short distance from an ocean port to a warehouse or other nearby destination is highly specialized. It is called "drayage," a historical term derived from the term "dray," a low cart without fixed sides suitable for carrying heavy loads a short distance.[25] The world of drayage is a world unto itself, with its own vocabulary and its own way of doing things.

A marine terminal, such as the APM Terminal at Port Elizabeth, is the place where containers imported by ocean carriers are transferred to inland carriers, including trucks, trains, and barges. Of course, for exports, it is the other way. The inland carriers bring the containers to the terminal. What mode of transportation is best for moving cargo in and out of New Jersey ports? Each mode has its own pros and cons. An advantage of the truck over the railroad is that it can provide door-to-door service. In other words, it is very flexible as to the precise delivery point. However, for longer distances, say 700 miles or more, the railroad begins to be a better deal. Rail transport is very good for providing connections between major cities, and rail transport is less affected by heavy rain, fog, or snow. Barge transportation has its advantages, with less air pollution than trucking or rail. It also consumes less energy per ton-mile of freight transportation, but its use in New Jersey and New York is rather limited. To keep things manageable, let us focus on the role of motor carriers, or trucks, which handle most of these containers.[26]

The trucking companies that bring shipping containers in and out of Port Newark–Elizabeth are highly specialized. They occupy a unique niche in the trucking universe. They are not long-haul truckers, and they are not exactly short-haul truckers either. Properly understood, they are *intermodal carriers*. They can take the containers from ships (one mode) and put them onboard a truck chassis (another mode), all the while not handling the goods themselves. Most of these truckers, about 70 percent, are private contractors. The advantage to the company in using private contractors is that the workload varies a great deal. If a private contractor elects to take a day off periodically, it is fine with the company. About 30 percent of the driers are company employees, but they are dedicated to a specialized type of cargo, where the workload is predictable.

Both types of trucks bring containers into the port for export, and they take imported containers to their final destination, often a warehouse. There are many such warehouses just a few miles south on the New Jersey Turnpike at Exit 8-A. Another final destination might be a big-box store such as Ikea. In any event, these intermodal carriers usually service an area within a 260-mile radius. Beyond that, it is cheaper for the shipper to choose another port or ship the goods by rail.[27]

To find out about these specialized truckers, I spoke with Dick Jones, who serves as the executive director of the Association of Bi-State Motor Carriers. It is an organization that advocates for some 170 intermodal trucking companies located in the New York–New Jersey area. They facilitate communication with the Port Authority as well as with state legislators, shipping lines, and terminal operators. The mission is to share ideas and to solve problems. According to their website, the Port of New York and New Jersey handles 80 percent of the world trade for a ten-state area, which represents 34 percent of the volume of trade for the United States.[28]

I asked Jones about the system of delivering cargo to the port. Each truck in the port is registered with the Port Authority and is given an identification number, or RFID. The RFID tag, mounted on the outside mirror on the driver's side, transmits information such as the booking number of the container to be exported, the cargo within the container, and the destination. The driver moves ahead and drives over the scale that captures the weight of the container as he moves along. It is important to know the weight of the container because there are weight limits for the cranes that put the container on the ship. In addition, the first mate on the ship needs to know the weight of each container so that the total load will be balanced. The machine in the entry passageway reads the RFID, calculates the weight, and quickly follows a computer program that decides where to place the incoming container, based on several variables, including most importantly the destinations.

As the driver approaches the exact spot where the container is to be dropped off, there is an RTG that can remove the container from the chassis and can place it on a stack. The weight of the container is a crucial variable in deciding where to place the container. Ideally, you want to place the lightest container on the ground. Then, you want to place the heaviest container on top of the stack because it will end up at the bottom of the ship. The most efficient arrangement is a stack of four containers high. Once you go over four containers high, you probably will have to handle that stack multiple times. It is possible to create a stack of up to seven containers high, but it is much less efficient. Time may be lost if the stack is too high, because it is likely to need to be rearranged.

The process for an intermodal driver to pick up imported container to be delivered to either a warehouse or an end user is much the same as delivering a container for export, with one big exception. When the driver leaves the terminal with the imported container, it is not weighed because the weight is no longer any concern of the terminal operator, who weigh containers only for export. The reason they weighed these containers was to determine their position aboard the ship, with consideration for balance and to certify that the weight limits for the cranes was not exceeded.

The imported containers were loaded out of the United States. Typically, the foreign shippers would just keep loading a container with as much cargo as possible. As long as they could close the doors, they kept loading. Then they would typically understate the weight of the container. There was no way for the trucker to know if the documentation was accurate. The overloaded container would be placed on a truck. Then the outbound intermodal driver would head down the road, and he might well be stopped at a routine checkpoint. There is always a chance of being stopped. The bill of lading might say that the container weighs 25,000 pounds, but once it is placed on the scale it is 55,000 pounds. Now, through no fault of his own, the intermodal trucker cannot run with the load because the police would impound the container. Now, there are fines to be paid by the trucking company. Moreover, the load can be delivered only by unpacking the container, lightening the load, and transferring the extra cargo to another, separate container. It was a big headache for intermodal carriers.

How to solve this problem? Fortunately, an effective leader stepped forward to find a resolution. This was Helen Delich Bentley, a colorful Maryland congressional representative, known for her fierce advocacy for the Port of Baltimore. Bentley realized that all these overloaded containers formed a tremendous problem, not just for Baltimore, but also for the entire industry. She had federal legislation enacted to make the shippers responsible for an accurate weight to be listed on the documentation. The legislation also gave the trucking company the right to hold the container, if they were caught with an overload, until the shipper paid the fines. The trucking company could get a lien against the cargo. The Bentley legislation made the shippers aware that they were responsible for accurate reporting of the weight of their containers.

It is nice when an elegant solution to a problem can be found as in the case of the overloaded foreign containers; however, some problems resist easy solutions. A persistent problem for intermodal carriers is terminal congestion that results in lengthy waiting times in long queues. Of course, no one likes wasting time waiting in line; but, for the intermodal driver, there is an additional problem. That problem centers on the electronic logging device (ELD), which is attached to the engine of the truck to record driving hours. The driving hours of truck drivers are regulated by rules known as Hours of Service (HOS). Previously, those hours were recorded in handwritten logbooks; however, today the books are all electronic. The drivers are working with a machine that is not flexible; it is all in black-and-white.

The truck drivers are allowed to drive eleven hours per day. In addition, they may have three more hours on duty, but not moving. At the end of those fourteen hours, the driver must go off duty for ten hours. Nearly everyone agrees that these rules are reasonable. The idea is to avoid accidents by preventing driver fatigue. However, there are wrinkles in the rules that work against the interests of the intermodal carriers. In the old days, with a paper log, if a driver were in a line at a pier, he could put down "on duty, but not driving." Now, if a driver is in line at a pier, he cannot log on as "not driving" because he has to keep moving, even though he actually is creeping along. If he goes more than five miles per hour, the computer records the truck as "driving." If a driver is in a slow-moving line at a pier, he has to move up every fifteen or twenty minutes. Meanwhile, he is using up his driving time.

As Dick Jones was explaining all this to me, I was quite astonished. I said, "It's unfair."

Jones replied, "But it's the law."[29]

As this is written, the law has not changed. If a driver moves his truck up a few feet while in a queue at a pier, he is "on duty while driving." It subtracts from his available time. Jones gave me an example of how the law works a hardship, "Let us assume a driver has a load from Port Elizabeth to Philadelphia."

In this example, the driver starts at six o'clock to pick up a load. He encounters an unplanned delay of four hours at the pier, and then he drives two hours to Philadelphia, where he encounters another unplanned delay of four hours to complete his delivery. Theoretically, he can get back as far as New Brunswick, where his tour of duty legally ends, and he cannot get home. His options are limited. If he continues on, he is liable for a fine. What the drivers want, of course, is an exception to the rules. When they are moving up in a queue, they want this time not be counted as "driving while on duty." It seems like a reasonable accommodation. The Association of Bi-State Motor Carriers and other industry organizations are lobbying for changes in the law, but it is a slow and expensive process.

Interestingly enough, not all of the imported containers are headed for a distribution warehouse or a big box store. U.S. Customs targets certain cargo for investigation. In these cases, the intermodal trucker will take the suspect container to a specialized warehouse. Customs maintains several locations in northern New Jersey. Such a place is called a CES (Centralized Examination Station). This is a privately operated facility designated by CBP for physical examination of cargo. The importer is responsible for all costs associated with a customs exam, including trucking to and from the CES, the CES fee, and storage.

Once the container arrives at the CES, it will be stripped of its contents. Customs inspectors will individually examine each carton, and then the cartons will be reloaded back into the container. Customs inspectors do the examinations, but they do not handle the cargo on and off the container. That work is outsourced by customs. They used to do it, but it takes too much labor. Contract workers open up the container and open up the boxes. Then a small number of inspectors carry out the examinations.

They are looking for discrepancies between the paperwork and the actual content of the containers. Of course, they are also looking for contraband. They have computer programs for targeting suspect containers. The container is registered when it is loaded in a foreign port. The selection for investigation is made while the container is en route. When the ship comes into port, the container is designated for CES examination. There are truckers who specialize in this niche business. Some five million containers come into Port Newark and Port Elizabeth every year, and only a tiny percentage can be examined.

Customs can use a high-tech examination called VACIS (Vehicle and Cargo Inspection System). They employ gamma ray technology to produce images of the container to look for contraband such as drugs, weapons, or currency. The system can also be used to inspect trucks and automobiles. They look for anomalies in the cargo, or strange shadows. For example, a container is supposed to have an automobile inside. Perhaps using VACIS, there is a cloud with some kind of a solid block. It does not match the profile of the documentation. Therefore, they open it up, and inspect it. A shipper who is caught several times is moved up to a "hot list." These are very serious federal crimes. Some have ingenious ways of handling cargo. For example, contraband can be stuffed behind the linings of the doors. Some place money or drugs underneath the floorboards. Others take out the fuel tanks and loads the fuel tanks with drugs. It is a constant game of cat and mouse. Some contraband will always get through. It is the job of customs to make smuggling as difficult as possible.[30]

As this is written, there is a looming controversy hanging over drayage trucking in New Jersey. How should owner-operators be classified? Should they be considered independent contractors or employees? The New Jersey Legislative Bill S863 addresses the possibility of reclassifying independent contractors in the state. Both sides have strong arguments. The Teamsters have been trying to organize drivers across the country. They support the employee model, arguing that it provides drivers with good wages and benefits, but unions cannot organize independent contractors. Thus, the Teamsters have been trying to get the legislature to classify drivers as employees. Meanwhile, there is strong objection to such a plan from most of the drayage companies in New Jersey. Based on long-standing practice, these drayage companies do business by charging a fee to either the cargo owners or the shipping lines for local haulage. Then they take a commission, and they pass most of the money along to the drivers. They argue that the drivers prefer to be their own boss and set their own schedules.[31]

In general, there has been strong support for the employee model from within American universities. Consider, as an example, the work of Michael Belzer, professor of economics at Wayne State University in Detroit, Michigan. He is an expert on the trucking industry and the author of *Sweatshops on Wheels: Winners and Losers in Trucking Deregulation* (2000). He has argued that the owner-operator model limits their bargaining power because they cannot take advantage of benefits such as workers compensation, unemployment, and retirement. He has argued that the industry would be better off if drivers were paid better: "The low road is not always the best road. If the industry accepted higher wages, it would stabilize the labor market, eliminate the 'driver shortage,' return better profits to companies, and improve service. It would allow the industry to escape this zero-sum game by reducing [driver] search costs and cut crash costs."[32]

To be sure, there are a few owners of trucking companies within the Port of New York and New Jersey that operate with the employee model. One of them is Tom Heimgartner, owner of Best Transportation in Newark,

"I compete every day as an employee-based company with independent-contractor-based companies. It's a little more expensive to do it our way. But I don't think it's going to be the end of the world."[33] The company's website explains that they are not worried about the impending threat of legislative reclassification. It reassures customers that there will be no interruption of service if the legislation becomes law. It says, in part, "Best Transportation does not use Owner Operators/Independent Contractors and never has. 100% of our drivers are company employees with full benefits and proper taxes paid.... We treat our team with the respect, compensation and benefits that they deserve. Our professional, tenured drivers average over 10+ years of experience each."[34]

On the other side of the argument, I have spoken with owners of independent-contractor companies who have said things, off the record, like, "I wish Tom Heimgartner well as an honorable competitor, and I defend his right to run his company his way. I just do not want the government to tell me how to run my business." This point of view is officially echoed on the record by Joni Casey, president of the Intermodal Association of North America. She has argued that the proposed legislation will cause a disruption in the intermodal supply chain, and it will increase shipping costs. She has said that state legislators simply do not understand how well the current system works for everyone, truckers and consumers alike. She has said, "These laws, which will affect several of the largest ports in North America, are about to restructure a linchpin of the intermodal freight supply chain without a clue to its impact on independent contractors or the American economy. They are not a good idea and will add unnecessary costs that ultimately will be passed on to consumers."[35]

When I discussed this issue with Philip Gigante of BBT Logistics in Newark, New Jersey, he gave me an eloquent defense of the contractor model. His company works with about forty truckers, all of whom are owner-operators. He argued persuasively that, since America was built on freedom, truckers should be free to choose whether they want to be contractors or employees. He said of his truckers, "We treat them as true contractors. They submit invoices every month or every week. We offer them a given task, which they can accept or reject. They alone determine where they want to go and when they want to go. They do make a decent living. Some of them have bought new trucks, and they have other drivers working for themselves. It sounds like a cliché, but it is the American dream for some of them. They come here to the United States, and they spend money on a truck. They make money, and then they are allowed to, just like me, invest in the business and buy another truck and then have their own employee driver."[36]

In approaching the trucking debate, I tried to take a fresh look at it, using a nonjudgmental, non-chip-on-the-shoulder approach. After talking with people on both sides, I found that the issue can serve as a kind of political litmus test. My colleagues at Rutgers are mostly liberals who tended to favor the expansion of government. My contacts at the Association of Bi-State Motor Carriers are mostly conservatives who tend to be suspicious of government. They would like to have government leave this decision to the private sector. I believe that there are many issues where the government should step in and regulate things, but intermodal trucking is not one of them. In January 2020, it looked for a while that the controversial bill might well pass. But then, there was some pushback because of things that were happening in California where a similar bill had been enacted. That bill had the unintended effect of badly hurting writers, musicians, translators, interpreters, photographers, and others working in the gig economy.[37] The law forced companies to hire these people as employees rather than as independent contractors. The result was that many companies simply refused and let them go.[38] Apparently the problems in California gave New Jersey legislators some pause. Then along came the COVID-19 pandemic, and everything was put on hold. In September 2020, the California legislature quietly revised the controversial bill with exemptions for many of those adversely effected.[39] As this book goes to press, it is difficult to know how this legislation will play out in New Jersey in the years ahead.

11

The Seamen's Church Institute

● ● ● ● ● ● ● ● ● ● ● ● ●

Coming from a long line of seafarers, and living in New Jersey, it was perhaps inevitable that I would become fascinated with the ports at Elizabeth and Newark. Driving north on the New Jersey Turnpike, I would often look over my right shoulder and see the giant cranes and the stacks of multicolored containers. I drove past the ports many times, and I kept saying to myself that I really should take a closer look. Driven by curiosity, I finally set aside a day to see these two ports for myself. But where to begin? I reached out to my colleagues at Williams College, a private liberal arts college in Williamstown, Massachusetts. I knew that they had an innovative Maritime Studies program that I admired. Sure enough, they gave me a good piece of advice. They suggested that I begin with a visit to the center for the Seamen's Church Institute (SCI) at 118 Export Street in Port Newark. In retrospect, that was, for sure, the best place to begin.

Armed with that advice, I set aside a day to explore the ports. I drove up the New Jersey Turnpike, and I took Exit 14, and I followed the signs for Port Newark. It is a place where few people visit unless they have some business there. I must confess that my first visit there was very intimidating. Driving in the port in my little car, I was dodging big tractor-trailers hauling multicolored containers with names like Maersk, Hanjin, Evergreen, and Overseas Orient Container Lines. Arriving at SCI, I introduced myself to the

FIGURE 31 The International Seafarers' Center of the Seamen's Church Institute at 118 Export Street in Port Newark. Opened in 1961, the center welcomes seafarers to a safe environment for relaxation and recreation. There they can use computers with good internet connections and use the money transfer service. (Photo taken on June 8, 2021, by Angus Kress Gillespie.)

director, the Reverend Jean R. Smith, and I explained to her my interest in the ports. She was most cordial and hospitable. She showed me around the center, and she explained what her agency does for seafarers and port workers—including truckers, stevedores, and warehouse workers.

Smith explained that modern seafarers often suffer from feelings of isolation and loneliness aboard ship. The hours are long, and there is little social cohesion onboard. The problem is compounded by the fact that the ships have quick turnarounds in port. So the work of the SCI in helping the seafarers during their brief time in port is crucial. There are practical services. The center offers them free Wi-Fi and the use of computers. Even more important, they offer international telephone service and low-cost phone cards, so that the seafarers can call home. In addition, they provide honest, low-cost money transfer service. Equally important, the ministers and the staff offer hospitality—a friendly face and someone to listen. Almost at once, I felt a deep appreciation for her work, and I scheduled a number of repeat visits.

In time, Reverend Smith encouraged me to accompany chaplains on ship visits to the ships in port. It was a wonderful opportunity. These ship visits were real eye-opening experiences for me. The chaplains would climb the long gangways up the side of the ship and be welcomed aboard, typically shown to the ship's galley, where they could meet with the ordinary seafarers. There were practical matters. The chaplains could offer phones, phone calls, SIM cards and "top ups." The chaplains could also offer money transfer services. Most importantly, the chaplains were good listeners. They could help with problems, and find solutions.

When the seafarers arrived in port, oftentimes the SCI would arrange a free shuttle service from the terminal gates to the nearby Jersey Gardens Outlet Mall with some two hundred stores, all under one roof. Typically, I would accompany the minister doing the driving there and back. On occasion, when they were shorthanded, they would let me drive the van. I remember vividly one day, while driving over to the mall, there was a young Filipino seafarer, in the back of the van, talking on his cell phone with his wife back in the Philippines. She was giving him her lengthy shopping list. I could not help but see the irony. He was about to shop for stuff that quite possibly he and his shipmates had brought over to the United States from China on a previous visit.

Over time, I found the port less intimidating and overwhelming. It was still exciting and interesting, but it gradually became familiar. It became clear that most all of the containerized cargo was arriving at the Elizabeth side of the port, while the Newark side was receiving all kinds of specialized bulk cargo. In all of these early explorations, I benefitted from my affiliation with the SCI clergy and staff members, who would explain the workings of the port to me. To tell the truth, my work on this book was enabled at every step of the way, thanks to the people at SCI for opening doors and introducing me to people.

We have explained throughout this book how New Jersey ports began to eclipse those in New York, and that story can be further illuminated by taking a quick look at the history of the SCI, an organization deeply rooted in Lower Manhattan. SCI got started in Lower Manhattan in 1843. There was a group of Episcopalian clergy, led by the Rector of Trinity Church on Wall Street, who wanted to establish a mission in what was called Sailor's Town. It was a rough and tumble area of Lower Manhattan with horrible conditions—particularly for seafarers. It was full of bars, gambling houses, and brothels, all designed to separate seafarers from their money. The system kept them in a constant state of debt to shipowners.

In that era of the 1830s, the ships were, of course, all sailing vessels that required huge numbers of crew members. The expansion of trade worldwide during that time required many sailing ships. There simply were not enough willing volunteers to crew the ships. So most seafarers were induced into the service through trickery. At that time, there was a system called crimping. Crimps were representatives of boarding houses. They would recruit seafarers to leave their jobs in the hope of getting a better job. Thus, the seafarers would forfeit all their pay from the original ship, and they would end up in debt to the boarding house. The crimps would then place them on another ship to collect a bonus for the placement.

The ships were so desperate for sailors that sometimes the crimps were actually putting dead people onto the ships. The crimps would douse the dead man with rum, so that he would appear to be in a drunken sleep. The shipowner would then pay the agent for this new crew member, only to realize later that he was not even alive. Many horrible things were going on down there in Lower Manhattan. It was in that environment that Episcopalians started a ministry among seafarers, motivated by altruism. The SCI became not only a place to provide worship but also an organization to advocate for social justice for seafarers.[1]

The first project of the newly formed organization was to build a floating chapel. The board purchased the ferryboat *Manhattan*, a catamaran or double-hulled boat of thirty feet in width and seventy feet in length. On top of this vessel, they constructed the Floating Church of Our Saviour for Seamen. It was a Gothic-style church with a seventy-foot steeple, spanning the twin hulls of the former ferryboat. After construction, it was towed to its home berth on the East River at the foot of Pike Street. Seafarers in attendance often prayed to God for deliverance from the dangers of the deep.[2]

Recognizing the need to go beyond the spiritual, the board decided to do more to meet the physical needs of the seafarers. We must remember that, back in the sailing ship days, it would take several days for a ship to unload its cargo and then several more day to load up with a new cargo. During that time, the sailors would be ashore, where they would be vulnerable to the temptations of wine, women, and song. The SCI wanted to provide for them a safer alternative. In 1850, they set up a lodging house called the Seamen's Home at 107 Greenwich Street. The facility could accommodate some five hundred men. It also provided a lending library and a rudimentary banking service to safeguard their earnings. Throughout the latter half of the nineteenth century, the SCI kept opening additional such land-based facilities as the need for their services grew.[3]

Reverend Archibald Romaine Mansfield was the superintendent of SCI from 1898 to 1934. At one point, he explained that the mission of the SCI was "to do for the sailor's good exactly what the crimps had been doing for his harm. They were the first to greet him in the harbour; so should we. They stood by him at the pay-off and took his money; so should we [for safekeeping]. They arranged for his food, lodging, clothing, and amusement; so should we. They provided him with a gathering place for companionship and social life; so should we. Finally, they got work for him when he signed off; *so should we.*"[4]

Under Mansfield's leadership, a new institute building at 25 South Street was opened on May 28, 1913. Part of the building was set aside as a hotel that could accommodate 580 men. Inexpensive dormitory beds were fifteen cents a night. Rooms that were somewhat more comfortable were twenty-five cents a night. On that very first night, ninety men took lodging there. The hotel provided mail services and a safe place for luggage storage. There were shower rooms with sinks for clothes washing, plus a drying closet. After washing up, the sailors could visit a lunch counter with good food at reasonable prices. There was even a very basic bank. It paid no interest; it was just for safekeeping. However, when a large sum was deposited, the sailor was advised to place the money in a regular bank that paid interest on deposits. The building was a great source of pride for the Reverend Archibald Romaine Mansfield.[5]

The building at 25 South Street was popular with seafarers in large part because the SCI placed its emphasis on practical help. To be sure, it was a religious organization, but the staff did not overtly proselytize. Prayer services were available, but they were not mandatory. As one old salt explained to young sailor, "You can eat there; sleep there; they'll stow your gear. It's cheap and it's clean. They got a job board—tell you what ships are lookin' for a crew. They got it all; and they ain't too holy."[6]

Over time, the thirteen-story building became a landmark in the downtown area because of a lighthouse beacon atop its roof. The beacon was officially a lighthouse that could be seen ten miles out to sea. The Coast Guard managed it. That lighthouse tower was a memorial to those who perished when the *Titanic* sank on April 15, 1912. The building worked well for the SCI for some fifty-two years; however, in June 1965, the board found that the building was obsolete, ant they decided to put up the building for sale at $2,500,000. It was not an easy decision because the building was home to a number of permanent residents. One of them was Sigurd Svendsen, age seventy-one, who had been there for five years, after spending forty-five years at sea as an able-bodied seaman under eight flags. Of course, he was unhappy about the looming closing of the building. He said that there was nothing else quite like it, because the it had "all the facilities."[7]

As we have seen, the 1960s were a time of tremendous change in the maritime industry. Containerization was rapidly replacing the old break-bulk methods. Interestingly enough, about the same time that that SCI was putting up for sale the obsolete building in Manhattan, they were putting the finishing touches on a new building in Port Newark. The building at 25 South Street was listed for sale in June 1965, and the new building at Port Newark was ready for dedication in November 1965.[8] With the wisdom of hindsight, we now can see that this would be continuing trend over the next fifty years or so: the maritime industry would shift from New York to New Jersey. To be sure, it did not happen all at once. The SCI had deep historic ties to Lower Manhattan.

After the sale of the building at 25 South Street, the SCI announced plans to build a new center at 9-19 State Street. It was to be $7,500,000 building, rising twenty-three stories above Battery Park. It was to have a five-story base with religious, recreational, and educational facilities. Above that was an eighteen-story tower for lodging some 340 seafarers.[9] The building was completed in 1967 at a cost of $9,700,000. Seafarers paid a modest rent to stay in the rooms, but the SCI ran the facility at a deficit. Containerization has sharply reduced the number of men needing lodging in Lower Manhattan, because most of the ships were over in New Jersey. Even there, the fast turn-around times meant that there was no real need for shore-side lodging.

By September 1985, just eighteen years after it was built, SCI sold the building. "Instead of being an asset, the building was becoming a financial drain," said Carlye Windley, a spokesperson. In addition, there was the cost of operating the building, built before the oil crisis of 1973, and lacking insulation.[10] At this point, the SCI moved its headquarters to temporary rented space at 50 Broadway, but the board was not happy with that location. They felt that the SCI had disappeared from public view, that it had become anonymous. After a three-year search, the board found a new location at 237–243 Water Street in the South Street Seaport Historic District in Lower Manhattan. Things moved quickly for the new $12 million building. Construction began in July 1989, and the dedication took place in May 1991. The six-story building used the façade of an eighteenth-century ship chandlery.[11]

The new building was rich in maritime decor. There were ship models of both sailing ships and power ships. There were ship prints and portraits of distinguished mariners, as well as an exhibit of traditional fancy ornament rope work, including such things as a hand boatswain's lanyard, decorative bell ropes, and the infamous cat o' nine tails. In addition, there was a Joseph Conrad room, with documents, letters, and photographs, honoring the English author of Polish descent, well known for his novel *Lord Jim* and his novella *Heart of Darkness*. There was also a small chapel with a font in the shape of a ship's capstan, originally used for hauling ropes and cables. Hung from the high ceiling was a ship model with female figurehead known as the Mystery Maiden.[12]

Significantly missing were any sleeping areas. The SCI was now clearly out of the hotel business. Containerization had changed everything. In its place was a Seafarer's Club with a snack bar, a reading room, and a counseling office. The beneficiaries of this club were mostly retired seafarers, who remembered the good old days with lengthy stopovers in Lower Manhattan. One of these old-timers was seventy-year-old Jim Lorier. He had first signed up as an ordinary seaman in 1942, and he had retired as a second mate in 1986. "Technology changes people's lives," he said. "You have fast turnarounds, ships in and out in 8 to 10 hours. I think the seaman's nature has changed. We used to have the camaraderie of three men sharing the same fo'c's'le. Now, there are individual rooms. I don't think anybody has to talk to anybody, except to argue over what TV program to tune in."[13]

For twenty years, this handsome building served as an iconic landmark for the SCI in Lower Manhattan, but then change caught up with it. In October 2010, the board placed the building on the market. Its eighteen employees were directed to report to the newly renovated center at Port Newark. Reverend David Rider, executive director of the institute, explained, "We're actually following exactly what the maritime industry has done. It used to be very much lodged in Manhattan, from the various shipping companies and bars and whatever else. But that's all gone. The punch line is containerization made it possible to move a box in a matter of minutes. So, these days, the seafarers coming in to see the world, they may not even get off the ship."[14]

Meanwhile, as the Water Street facility in Lower Manhattan was shutting down, the Port Newark center was busier than ever, serving some 15,500 sailors and more than 400,000 truckers annually. One of that center's regular visitors was Nick LeBlanche, the first officer of the M/V *Oleander,* a relatively small container ship at 387 feet in length. This ship was a regular feature at Port Newark, taking supplies to Bermuda one week, and returning the following week with deliveries of fresh fish and dry goods. He was particularly grateful for the services rendered by the SCI for his crew of fifteen men: "Without them, the guys are not able to go out and do some shopping because basically, each week, they provide the transportation service, and the money exchange for their families—and without them it would not be possible to do that. Eventually, it would be a problem if the church was not there because the rules are complicated."[15]

FIGURE 32 The Reverend David Rider served as president and executive director of the Seamen's Church Institute from 2007 to 2020. During this period, Rider implemented a comprehensive upgrade of the Port Newark International Seafarers' Center that provides hospitality to thousands of seafarers arriving there every year. (Photo taken on May 5, 2016, on the roof of the Seafarers' Center. Photo credit: Leo Sorel.)

Over the years, the mission of the SCI has changed along with the times. The maritime industry has moved out of Lower Manhattan. The container ships are now docking in New Jersey. Merchant ships simply do not come to Manhattan anymore.[16] SCI had this beautiful signature building at Water Street on very valuable property that was not being put to its highest and best use. The board questioned whether it was wise to have all these resources used to maintain the structure that was no longer relevant to the actual work of the SCI. Active seafarers never visited that building. There were no ships nearby.

The decision was made to keep a small rental footprint for the SCI national headquarters in Lower Manhattan, but the real work of the East Coast ministry was in Port Newark and Port Elizabeth.[17] The board decided to sell the

Water Street building and to use some of the proceeds to modernize and upgrade the building in Port Newark, to make it relevant to the needs of the present. Today, it is the most handsome building in the whole port region. When seafarers come into the center, they can get on their computers with free Wi-Fi. Though most of them have their own computers, the center has some if they do not. They can get on Skype and talk with their families back home and make a connection. It is their place.[18]

In the old days, seafarers used to spend several days in port; now they often only have a few hours. Moreover, many seafarers arriving in U.S. ports—even if they have time—do not have visas permitting them to leave their ships, thus limiting their access to services on land. As a result, some of these mariners must postpone much of what we take for granted—emails, phone calls, banking—until the next port of call in a different country, which could take weeks to reach. However, because SCI ship visitors board each vessel as it arrives, seafarers can connect with friends and family following a long ocean voyage.[19] It was my good fortune to accompany the Reverend Megan Sanders on several of the ship visits.

I spoke with the Reverend Megan Sanders, thirty-five, as she went about her pastoral duties, driving a large, blue, twelve-passenger van with the SCI logo with cross and anchor painted on both sides, around the port, transporting seafarers and visiting ships. An only child, her father had been a Green Beret in the Vietnam War, and her mother was a civil servant in the National Security Agency (NSA), an intelligence agency of the U.S. Department of Defense. Her parents split up when she was only two years old, so she was raised almost exclusively by her mother, who lived in Northeast Washington, D.C., only two and a half blocks from the White House. When Megan was twelve years old, her mother was transferred to Hurlburt Field, a U.S. Air Force installation in the Florida Panhandle.

Megan was brought up in the Catholic religion, and she attended Flagler College in St. Augustine, Florida. Going through college, she pretty much decided that she wanted to enter the priesthood, but being female and Catholic, that door was closed to her. The day may come when the Catholic Church agrees to ordain women, but Megan could not wait. She decided to switch to the Protestant Episcopal Church, and she enrolled in the General Theological Seminary in the Chelsea neighborhood of Manhattan. Upon graduation, she served as assistant pastor for the Episcopal Church in Essex Fells, New Jersey. After a year and a half there, she found work at SCI in Port Newark, with the aid of a computer matching system that connects ministers needing work with institutions needing ministers. When I spoke with her, she had already been working at SCI for some four years.

Driving around the port, Megan is careful to stay in her clerical role and to minimize her feminine side. Wearing little or no makeup, she fits into the hypermasculine, blue-collar, backslapping atmosphere of the port. Otherwise, she would be fending off unwanted advances day after day. The port is a place for hardworking, manly men with swagger. There is no place for the weak, the timid, or the frightened. It is a place for big men who operate big machines. Meeting up with these men—the truckers, the security guards, and the longshoremen—I was reminded of the fictional character Flex Crush, a 225-pound truck driver, who made his living hauling loads of nuclear waste. He was a guy whose breakfast was made up of "steak, prime rib, six eggs and a loaf of toast." This was eaten with side orders of flapjacks, a pound of bacon, and a pound of roofing nails. As he ate, he would be "idly cleaning his 12-gauge shotgun."[20]

The chaplains at SCI have two main areas of responsibility. Their regular parishioners, so to speak, are the port workers—the truck drivers and the warehouse workers and the food truck operators and the security guards and the longshoremen. One might say that the seafarers make up their mission congregations, though they make a point of not evangelizing. They do not preach or try to convert anyone. They will offer prayers and blessings, if asked, but they do not force themselves on anyone.

Early that morning, the dispatcher sent Megan to pick up four Filipino seafarers from a ship and take them to Jersey Gardens, a nearby two-level indoor outlet mall in Elizabeth. It happens to be the largest outlet mall in New Jersey, built on a former landfill. The Victoria's Secret store is especially popular with seafarers who like to buy perfume and seductive clothing for their wives and girlfriends. The mall seems to have just about everything. The stores include Bed, Bath & Beyond, Calvin Klein, the Disney Store, Lord & Taylor, Nike, and so on. Strangely enough, the mall does not sell electronic goods, very much wanted by seafarers. For these, they must go to Walmart or Best Buy in Union, New Jersey.

The ship was docked within the PNCT (Port Newark Container Terminal) portion of the port. However, given the tight port security, we could not just drive up to the side of the ship. We had to drive up to the gate and present our TWIC (Transport Workers Identification Credential) cards to the security guard, who opened the gate arm. Then we had to pull over and await a security escort in van with a flashing yellow light. After a short wait, we followed the escort through a maze of concrete barriers and idling tractor-trailers waiting for their loads. We picked up the seafarers and took them to mall, about five minutes south. As we dropped the seafarers off at the mall entrance, they waved goodbye when we pulled away.

"I sometimes feel like a soccer mom, driving these guys around," said Megan, with a gentle smile. "I feel blessed helping these invisible people, set apart from mainstream society. You are injected into their lives. I ask them about their families and their children. Even though they have a hard life, these are the lucky ones. They are able to send their children to private school and their siblings to college."[21]

Later that afternoon, we called upon the MSC *Silvana*, a large container ship operated by MSC (Mediterranean Shipping Company), a Geneva-based company operating in all major ports of the world. MSC is the world's largest shipping line in terms of container vessel capacity. The *Silvana*, at 1,089 feet long and 141 feet wide, is truly huge. We parked next to the ship. The ship had arrived at seven o'clock, the morning before. It was now getting ready to leave at four o'clock that afternoon, a typical quick turnaround. The simultaneous unloading and loading of containers had been completed. On the pier side, a ship's chandler was loading food and supplies for the continuing voyage. On the channel side, a barge was pumping on bunker fuel. It looked like a mountain as we climbed the fifty-five steps along the aluminum gangway, holding onto the ropes on either side. When we reached the top, we spotted a large sign: "WARNING: NO OPEN LIGHTS, NO SMOKING, NO VISITORS." We were greeted by a crew member who took down our names and time of arrival in logbook.

We were escorted to the ship's office where Darko Gardic, the second officer, who was on duty, received us. Tall and thin, he was wearing dark black coveralls with the MSC logo. Very hospitable, he offered us coffee. In turn, Megan offered the usual services of SCI, including the sale of phone cards and the offer of transportation to nearby attractions. Meanwhile, I asked Darko Gardic about his background and current duties. He was from Montenegro in Southeastern Europe, where he had graduated from the maritime academy there. After many years at sea, he had worked his way up to the rank of second mate, a position that placed him third in command, after the master and the chief mate. I asked him about his duties at sea. He explained that he worked on the ship's navigational bridge, filled with sophisticated equipment. On a typical day at sea, Darko would work from noon to four in the afternoon, and again from midnight to four in the morning.

He told me, "This is not a nine-to-five job. It's very tiring, and you are only paid when you are working. In a typical year, I spend nine months working and three months at home. It's a dog's life. Only people from poorer countries like Montenegro are willing to do it," he said laughing.

I asked him why he was laughing. He said, "I have to laugh to keep from crying. But what choice do I have? I have to sacrifice for my family, my wife and my eight-month-old daughter. I hope to have more kids, enough one day to make up a whole soccer team."[22]

Later that afternoon, we visited the container ship *Wan Hai 501*, out of Singapore, with a crew of nineteen. The ship had arrived at 06:00, the day before, and it was scheduled to leave at 16:30 that afternoon. The ship had come from Shanghai, China, through the Panama Canal to Port Newark. It was then scheduled to continue on to Norfolk and Savannah before returning to China, completing a three-month circuit. We were received in the wardroom by Second Officer Yahya Musbaiton from Indonesia, handsome in khaki uniform with almost movie-star good looks. In port, his duty hours were noon to six in the evening and again from midnight to six in the morning; in other words, he was on duty twelve out of every twenty-four hours. On the bulkhead, or wall, of the wardroom were various orders and placards including one that said, "Don't throw garbage overboard. It's against the law."

On this ship, the master was Taiwanese and the first officer was Chinese. I asked Yahya about his background. He told me that his father had been an officer in the Indonesian Navy and that his father had sent him away to a maritime academy when he was fourteen years old. After many years at sea, he had worked his way up to the rank of second officer.

I asked him about life at sea. He said, "It's a hard life. I am away from home too much of the time. I missed the funeral of my father and the funeral of my mother. I have a three-year-old son and six-year-old daughter. I have missed most of their growing up. But I have no choice. Coming from Indonesia, my seaman's salary is relatively high. I make as much in one month as ship's officer as I would in two years in regular job back home."

I noticed that Yahya was wearing the standard open-collar khaki uniform shirt with loops for shoulder boards, almost universal among merchant marine officers, yet he was not wearing shoulder boards with gold stripes. He said, "Well, I want the men to respect me for my knowledge and my position, not for my uniform." I asked him how much longer he would continue to serve at sea. He said, "I hope someday to be able to teach in one of the maritime academies back home. It would be a much better life."[23]

After we left the ship, I asked Megan how she was able to relate to different seafarers, day after day. After all, it is hard to minister to someone who is only in port for a few hours. Megan said, "I grew up as a transient person. If you move around a lot, you have to build community quickly. It's a gift that was given to me by my family that enables me to build community on these ships.

"These seafarers do not have a geographical anchor. We try to anchor them to God—as they understand God. Then, they are home, at least spiritually. They feel less isolated through their conversations with chaplains.

"I think of myself as a Spiritual Skycap. I am only with these seafarers for a short time, but I help them carry their baggage. I help them lay down their baggage."[24]

A week later, I accompanied Megan on her visit to the CMA CGM *Tosca*, which was berthed at the APM Terminal in Port Elizabeth. The ship was of French registry with its home port in Marseille. About half of the crew, including all of the officers, was French. The rest were either Romanian or Asian Indian. The terminal is vast, with some 350 acres, fifteen large cranes, 6,001 feet of wharf, three deep-water berths, and on-dock rail service. We drove up to the gate in a blue SCI passenger van, but you cannot just drive in to meet a given ship. You must first wait for an escort van with a flashing yellow light. While waiting, I asked the security guard about the meaning of the initials "APM." He explained that it stood for Arnold Peter Møller, a Danish shipping magnate who was the founder of the A.P. Moller-Maersk Group back in 1904.

Waiting for the security escort can be time-consuming and frustrating. It took a full forty-five minutes for the security van to arrive. That gave me plenty of time to study the various precautionary signs posted at the gate. One prominent sign to the right of the gate said, "NO VEST, NO ENTRY." Another sign said, "PROTECTED: No Private Vehicles Permitted in Yard." Still another said, "NO SNEAKERS PERMITTED, MUST WEAR PPE."

"What's PPE?" I asked Megan. "That's Proper Protective Equipment," she explained.[25]

The chain-link fence was festooned with such welcoming signs. It made for interesting reading. A sign to the left of the gate had this stern warning: "Secure Area: TWIC Required. Non-TWIC Credentialed Personnel Must Be Escorted." Fortunately, I had my TWIC card around my neck. Because I had to jump through many hoops to get it, I knew that it stood for "Transportation Worker Identification Credential." I took good care of that card since it provided access to secure areas of the port. In addition, it cost me $129.75, an awful lot of money for an ID card. However, I was in good company. Other cardholders include Coast Guard–credentialed merchant mariners, port facility employees, longshoremen, and truck drivers.

Of course, even though it seemed to take forever for our security escort to arrive, loitering at the gate is not allowed, as another sign warned: "RESTRICTED AREA: Authorized Personnel Only. Unauthorized Presence Constitutes a Breach of Security." I was fully prepared to explain to the police that I was not loitering; I was just waiting for an escort.

Finally, the security van arrived, and we pulled up alongside the CMA CGM *Tosca*. At more than a thousand feet in length, it was one of the biggest container ships I had ever seen. We were escorted up to the ship's office, where we were greeted by First Officer Blot Guilhem, who offered us coffee. Over the coffee serving area was posted a sign, "Merci de Laisser cet Endroit Propre," or "Please Leave this Area in Good Condition." Over coffee, I asked Second Officer Jean-Marie Franceshi, "How did the ship get its name?" He patiently explained that it comes from *Tosca*, an opera in three acts by Giacomo Puccini. It premiered at the Teatro Constanzi in Rome in 1900. I was embarrassed that I knew so little about opera.[26]

Since the ship did not leave until nine o'clock that night, there was adequate time for select crew members to take a bit of shore leave, and we offered to transport them to either the Seamen's Center or to Jersey Gardens. There were seven crew members who wished to go ashore, but they needed to take their lunch first. So, as luck would have it, they invited us to join them for lunch. It was all very pleasant. It was a white-tablecloth luncheon. We were served steak smothered in mushrooms with a side of mixed vegetables along with freshly baked bread. For dessert, there was a cheese tray served with fresh fruit. In typical French fashion, no one rushed through the meal. Instead, everyone took their time and the meal was punctuated with lively conversation. Though my French was severely limited, they made an effort to include me.

After lunch, we descended the gangway and took five seafarers to Jersey Gardens for shopping, and two officers to the Seamen's Center. It turned out that the two officers were not interested in shopping; instead, they were eager to take advantage of the internet connection provided at the center. In conversation with the crew, I had caught bits and pieces of the route of the ship, but I did not have a clear picture. So as we drove toward the Seamen's Center, I asked First Officer Blot Guilhem to spell it out for me. He carefully explained that, after Port Elizabeth, the ship would go south to Norfolk and Savannah. That would wrap up the tour of the U.S. East Coast, and the ship would head east across the Atlantic Ocean, pass through the narrow Straights of Gibraltar between Morocco and Spain, and enter the Mediterranean Sea. The ship then would continue east, enter the Suez Canal at Port

Said, exit at Suez, pass through the Red Sea and through the Gulf of Aden, rounding the horn of Africa. Now off the coast of Somalia, the ship would enter the Indian Ocean—clearly the most dangerous portion of the journey because of the threat of piracy.

Although this area is indeed very dangerous, the CMA CGM *Tosca* does not carry any armed guards. Instead, the crew relies on staying within the Maritime Security Patrol Area, a narrow corridor through the center of the gulf that has been patrolled by an international group of warships since 2008. They take the additional precaution of putting up strands of razor wire across the stern, since this is the lowest and usual point of entry for pirates. Though the normal economical cruising speed for the ship might be 12 knots, here the crew cranks it up to 18 knots or more in order to outrun the pirates. Oddly enough, the crew welcomes bad weather and rough seas here, since it keeps the pirates ashore.

Continuing east, the ship would round the subcontinent of India and proceed to the first Asian port—Tanjung Pelepas in Malaysia. This container port, only opened in 1999, has already set a world record as the fastest growing port. Its amazing growth can be explained by its proximity to the busy sea lanes that serve Singapore. In fact, Tanjung Pelepas now provides shippers an attractive alternative to its competitor, Singapore. The next stop for the CMA CGM *Tosca* would be the deep natural harbor of Hong Kong, a special administrative region (SAR) of the People's Republic of China, situated on China's south coast. After Hong Kong, the ship would stop at three separate Chinese ports—Yantian, Shanghai, and Ningbo. The last stop in Asia would be the port of Busan, formerly spelled Pusan, South Korea's second largest metropolis after Seoul.

The ship then would head east across the Pacific Ocean to the coastal seaport of Seattle, Washington. As this was written, Seattle was the third busiest container port in the United States, after Los Angeles–Long Beach and New York–New Jersey. The next stop would be Vancouver, a coastal seaport city on the mainland of British Columbia, Canada. It is the largest port by tonnage in Canada. Then the ship would head west, back across the Pacific to Asia. The first Asian stop on the return leg of the voyage would be Yokohama, one of the major ports in Japan. The ship would then repeat some of the ports while continuing west—stopping at Shanghai, Ningbo, Yantian, and Tanjung Pelepas—on the way to the Suez Canal and west through the Mediterranean and across the Atlantic, back to Port Elizabeth.

It is interesting to note that the voyage originates on the East Coast of North America and that the turnaround point is on the West Coast of North

America. Why not just travel from Vancouver to Port Elizabeth by way of the Panama Canal, making a circuit of the globe, rather than going back and forth across two oceans? The answer, as this was written, was that the ship was simply too wide to fit into the locks of the Panama Canal. The CMA CGM *Tosca* has a width of 42 meters, but the Panama Canal then allowed for a maximum beam of only 32.31 meters.

The entire circuit takes three months and three weeks, or about 113 days, just like clockwork, or much like the predictable route of a city bus. There are some noteworthy aspects to all this. One is that the shipping company, in its official publications, refers to Port Elizabeth in New Jersey simply as "New York, US." Although this may hurt the pride of New Jerseyans, myself included, it makes sense in the global picture, where New Jersey is just seen as a small part of the greater New York area. The alert reader may have also noticed that the CMA CGM *Tosca* flies the French flag and is manned by French officers, but it makes no stops in France. In fact, it makes no stops anywhere in Europe. The main purpose of this vessel is to ship goods made in Asia to consumers in North America.

Meanwhile, it should be noted that the basic contract for the French officers calls for them to work three months at sea, followed by three months of shore leave at home. Their rotation is generous when compared with that of third-world seafarers. It is clear from the above itinerary that they cannot get on and off the ship in the supposed home port of record in Marseille, France, since the ship never goes there. Typically, they report for duty by flying from France to New York, where they board the ship. And the reverse is true. When their time onboard the ship is up, they fly from New York back to France.[27]

We arrived at the Seamen's Center, and Megan parked the van. I walked into the lobby of the building, with the two French officers. First Officer Blot Guilhem, who had clearly been here before, was immediately drawn to the large, brass ship's bell on prominent display in the lobby. He explained that the bell had been salvaged from the SS *Normandie*, an ocean liner built in France for the French Line Compagnie Générale Transatlantique. When she entered service in 1935, she was the largest and fastest passenger ship afloat. When World War II broke out, the U.S. government seized the Normandie. In the process of being converted to a troopship, the liner caught fire, capsized, and sunk into the mud of the Hudson River at Pier 88 on the West Side of Manhattan. The damage was extensive, and it was decided that restoration would have been too expensive. The ship was scrapped in October 1946. Fortunately, some of the artifacts, including the bell, were salvaged.[28]

FIGURE 33 In the lobby of the Seafarers' Center, we find the salvaged ship's bell from the French ocean liner SS *Normandie*, which caught fire during World War II while docked at Pier 88 in New York City. The ship's bell was used for traditional functions, and it has the name of the ship engraved on it. (Photo taken on June 8, 2021, by Angus Kress Gillespie.)

At the end of the day, back in the chaplain's office, I asked the Reverend Megan Sanders how working as a port chaplain differs from being a regular parish priest. She explained, "Here at the port, I am doing meaningful outreach every day. It's a kind of 'rubber meets the road' ministry. On the other hand, in a regular Episcopal parish, the priest is under the direct control of the vestry, the elected governing body of the parish. Since the vestry has the power of the purse, they feel that the priest must answer to them. Of course, here in the port, we must ultimately answer to the board of directors. But, on a day to day basis, we pretty much operate independently, without parish politics. We get to do our work without being micro-managed."[29]

Later, I stopped by the International Seafarers' Center to catch up with how things had changed during the COVID-19 pandemic. I spoke with the Reverend James Kollin, the most senior chaplain at Port Newark, having served there for some twenty years. He grew up in Benguet in the northern Philippines. He studied divinity at the St. Andrew's Theological Seminary in Quezon City, and he graduated in 1987. After graduation, he worked with a church in the Philippines for some ten years. At that point, he was looking for a change, so he approached his bishop to ask about additional opportunities. The bishop then recommended Kollin for an internship with the SCI at Port Newark, where he served in 1998 and 1999. After completing the internship, he returned to the Philippines to do similar work there. After about a year, he got an email asking if he would like to be a full-time chaplain back at Port Newark, which he has done ever since.

I asked Kollin what his work was like in normal times. He explained that it's all about ship visits. The several chaplains report to the center in the morning, and they go over the list of ships that are currently docked at the port. They equally distribute the number of ships to the number of chaplains. Each chaplain gets into his or her own designated van and heads out to the port. The chaplain goes up the gangway and meets with individual seafarers in need. A crew member might need to talk about a family matter, a sickness, a death, or relationship issue. The chaplain can offer listening, support, counseling, and prayers. The key thing is that when a crew member sees a chaplain, there is an element of trust and confidence, so the seafarers feel free to express whatever is on their mind. In addition to counseling, the chaplain would often offer to use the van to take a group of seafarers from the ship over to the Jersey Gardens shopping center and back.

I asked Kollin how things had changed during the COVID-19 pandemic. He explained that the most serious issue was that crews around the world were not being routinely rotated. Normally, a merchant seafarer would sign a six- or twelve-month contract. At the end of the contract, the seafarer might get off the ship at some foreign port and get a charter flight to his or her home country. Meanwhile, a replacement would arrive at the ship.

Global travel restrictions changed all that. It was difficult to fly in replacements. As a result, the existing crew began to feel trapped, almost imprisoned, and unable to get off the ship for well over a year. Even when ships came into Port Newark, they could not get shore leave. They could not visit the Seamen's Center, nor could they go shopping at nearby Jersey Gardens. How did the SCI respond to this humanitarian crisis? They developed something called shop-at-sea. In brief, they developed an online ordering service.

Seafarers who knew that their ship was headed for Port Newark or Port Elizabeth were encouraged to place their shopping orders for essential items two weeks in advance of their arrival. They could select items from Target or Best Buy and email their request to SCI, which would go ahead and order the items and have them delivered to SCI, where they would await the arrival of the ship. The program was limited to essential items such as medications, clothing, snacks, and necessary electronic equipment. Each crew member had a limit of five items. Heavy and oversized items such as a bicycle or a 72-inch television were excluded.[30]

I accompanied Kollin on a ship visit to the SC *Marigot*, a container ship on June 8, 2021. We loaded up a van with three large blue plastic bags full of the crew's items, and we drove south to the Maher Terminal, where the ship was docked. We climbed up the gangway with the shopping bags, and we were escorted up to the ship's office where crew members were summoned, one at a time, to receive their orders. Careful records were kept. Each crew member paid cash to reimburse SCI for the order.

I got the sense that these were not just commercial transactions. The chaplain was doing a favor for each crew member as pleasantries were exchanged. It was a bright spot in a special day—a brief relief from boredom and isolation. Here was a message that somebody cares about you. I felt honored and pleased to have been given the chance to share these moments.

Later, I had the opportunity to meet with the newest and youngest chaplain at the Port Newark SCI, the Reverend Cora DiDomenico. She grew up in Chicago, where she told me she was familiar with trucks and trains, but not with ships. Later, after college, she studied at the Union Theological Seminary in New York City. There she studied psychology in religion, with a focus on grief studies, and she graduated with a master's degree in divinity in 2018. While still studying at Union Theological, she took an internship as a youth pastor at a parish in New Jersey. She told me that it was not a good

FIGURE 34 After delivering the requested shop-at-sea items to the crewmembers, Rev. James Kollin (left) meets with Captain Francisco S. Baguio Jr. of the SC *Marigot* to discuss the pastoral counseling and support needs of the crew. Because of the COVID-19 pandemic, many of the crew were suffering from feelings of isolation and fatigue because of restrictions on crew changes and the lack of shore leave. (Photo taken on June 8, 2021, by Angus Kress Gillespie.)

fit, so she consulted with the placement office at Union Theological about finding something different. They knew that she was living in New Jersey and that she had a car, so they suggested that she take an internship with SCI at Port Newark. For students living in New York, it was a difficult commute; but, for Cora, it was quite workable. She started interning in 2017; and, when she graduated in 2018, she was brought on full-time as a chaplain.

I asked DiDomenico what the work was like for the first couple of years before the pandemic outbreak. She explained that everything was always centered on having meals with the seafarers when she would board the ship. It might mean sharing coffee at midmorning or taking lunch at noon or sharing coffee at midafternoon, depending on the schedule. It was all about sharing and being together. It was in the seafarer culture to provide hospitality to visitors. At the same time, the chaplain was offering different services such as sending money with SIM cards or providing pastoral care. It was almost always around the table, around food. At the same time, a chaplain might provide transportation to the nearby shopping mall. While driving those crew members places, it was often possible to have conversations and figure out their needs.

I asked DiDomenico if she had ever gone the extra mile for someone. She shared with me a moving story about a seafarer on a tanker whose glasses broke. He was a fitter in the engine room, and he wore a welding mask to make new parts. Without glasses, he could not see to do his job. He was from the Philippines, and so he had a Filipino prescription, which did not compute to an American prescription. Cora was determined to help, but she worried that eyewear in the United States would be too costly for a Filipino. Fortunately, his girlfriend sent him some money. Cora took him to Walmart because that was the least expensive eye exam that she could find. She then took him to LensCrafters, where they made the glasses in two hours. There they caught a lucky break. The technician there came from an American Merchant Marine family, and he sympathized with the distraught seafarer. He gave the Filipino an employee discount, so the $700 glasses ended up costing about $250. Cora worked with the seafarer from about three o'clock to nearly nine, but she got him the glasses. It was lucky that the seafarer was from a tanker because that meant that he had more time in port. Had he been from a container ship, it might not have worked because of a short stay. The ship sailed the next day.

On another occasion, DiDomenico visited a container ship where a senior crew member had died of COVID-19 after being in South Africa. The crew did not know what to do, but they reached out to SCI first, even before the Coast Guard. Many of the seafarers on that vessel were also sick, but they did not know if it was a result of COVID-19. They thought that perhaps they were not well because they were not sleeping enough and they were working all the time.

Cora went ahead and boarded the vessel, and she prayed over the body of the man that had died. She then spent a full fifteen hours with the crew because the man who died had been like a father figure for them. She provided a great deal of grief counseling for them. Many of them were not sleeping well. They were sleeping together in one cabin because they were anxious and unable to sleep alone. Cora got them headphones so they could listen to music to help them go to sleep. She advised them to turn the lights off when sleeping so as to get better rest. Later, she shared a meal with them.

Afterward, at first, DiDomenico was angry that she had been exposed to COVID-19. But upon reflection, she realized that, because of her services, the ship was able to sail and continue. Without her, the crew might have given up and gone home. There was plenty of time to mull over the experience as she had to quarantine for two weeks afterward. In the end, she was relieved that kept getting negative test results.

FIGURE 35 Chaplain Cora DiDomenico leaves the Seafarers' Center on June 8, 2021, for Jersey City to shop at Best Buy and Target for items like toothpaste and electronics for the crew of her next ship visit at the CMA CGM *Argentina*, a container ship that was built in 2019 and is sailing under the flag of Malta. This kind of personal shopping is necessary because the crew cannot leave their ship. (Photo by Angus Kress Gillespie.)

Finally, I asked DiDomenico about the future of this ministry in the postpandemic world. She raised an issue that, honestly, I had not considered. She explained that, in the prepandemic era, seafarers' contracts were getting shorter for reasons of their mental health. It is not a good idea to have someone on board a vessel for a nine-month period. It is far better to have them for a six-month period. But swapping out the crew every six months starts to get costly. With the pandemic, some people were on board for twelve months, even seventeen months. They were trapped. Cora said, "I worry that companies are going to take advantage of that and go back to being able to use people for longer periods of time."[31]

12

The Future

● ● ● ● ● ● ● ● ● ● ● ● ●

To properly tell the story of the future of the global supply chain in all of its complexity, we should examine all of its component parts including ports, ships, trucks, railroads, warehouse, and much more. However, to keep our task manageable, let us focus on just two key components—ports and ships. Let us start with ports, specifically Port Newark–Elizabeth. Our port has some distinctive and unique features. To be sure, these include a favorable location with access to a large consumer market and strong governmental support in having dredged deeper channels and having the bridge spanning the Kill van Kull raised to accommodate larger ships. At the same time, looking at the future, our port is exposed to the same risks and it has access to the same opportunities as other ports worldwide.

With the largest collection of maritime cargo handling facilities on the East Coast of North America, the future of Port Newark–Elizabeth seems bright. This 2,100-acre complex, located on the eastern shore of Newark Bay, has a well-established range of maritime commerce activities including container handling terminals, automobile processing facilities, break-bulk facilities, trucking firms, and an on-dock rail terminal. There are piles of rock salt, huge mounds of Belgian blocks, not to mention giant tanks of orange juice. As a key component of the larger Port of New York and New Jersey, the port has access to more than 125 million consumers in the Northeast, Midwest, and Canada by truck or rail. There are one billion square feet dedicated to warehousing and distribution in the region.[1]

With a good location and heavy investment and knowledgeable people, it might seem that the future of the port is assured.[2] However, as the old saying goes, we plan, God laughs.[3] I find writing about the future of the port to be a rather frightening challenge. It is a more difficult task than describing what has happened in the past. Others before me have faced the same difficulty and have warned against being too bold. I am reminded of the advice given by Winston Churchill: "It is a mistake to look too far ahead. The chain of destiny can only be grasped one link at a time."[4] With that advice in mind, I plan a twofold approach to the future of Port Newark–Elizabeth. In this chapter, I will first discuss both likely and possible disruptions, followed by a discussion of the opportunities that lie ahead. Both phases are full of unknowns and pitfalls, but I will give it my best shot.

Port Disruptions

There are plenty of potential disruptions to the predictable and smooth running of the port that lie ahead. Some of the disruptions are predictable. These include familiar things like storms with flooding and the occasional power outage, as well as labor disputes from time to time. Other disruptions take us by surprise, such as COVID-19, which led to supply chains disrupted by factory shutdowns, expensive megaships sitting idle, and cargo terminals everywhere in the United States suffering from a fall-off of imports.[5] Port employers began asking federal and state governments for assistance to cover the cost of safety measures aimed at keeping terminals open. John Nardi, the president of the New York Shipping Association (NYSA), wrote that the safety measures "create extraordinary costs which, in the absence of state and federal assistance, will have negative impacts in the near future." He was referring in part to the fact that the NYSA was spending $60,000 weekly on daily temperature checks. Meanwhile, marine terminals and the International Longshoremen's Association (ILA) was handing out over 200,000 masks and barrels of hand sanitizer.[6]

To be sure, the sudden outbreak of an infectious disease that spreads rapidly to many people can justifiably be called an unexpected disruption, taking policy makers by surprise. On the other hand, we have had experience with other port disruptions that were entirely avoidable brought upon us by poor decisions by policy makers. An example is the recent U.S.-China trade war initiated by President Donald Trump, who fatefully said in March 2018, "Trade wars are good, and easy to win."[7] The data from the following year, clearly showed that the United States had not won anything. U.S. exports to China registered at $8.7 billion in July 2019, down from $10.1 billion in July 2018. U.S. exports were down in agricultural products including cotton, tobacco, dairy products, livestock, and meats. To be fair to Trump, all the experts agree that there were real problems with the U.S.-China trade that needed be addressed. It's just that tariffs were not the solution. As Jonathan Gold, president of supply chain and customs policy at the National Retail Federation, said, "There is no doubt that China has been a bad actor. There are serious issues to be resolved on intellectual property rights, forced technology transfer (and so forth), but tariffs are not the right way to get there. We need to work with all our allies instead of using the heavy-handed tariff approach."[8] What politicians like Trump tend to forget is that, in a trade war, just as in a real war, the enemy gets a vote. Hopefully, future administrations will seek to avoid unnecessary trade wars.

What could be more frightening than a botched trade war? How about a malicious cyberattack? We got an idea the seriousness of this problem back in June 27, 2017. At that time, there was a ransomware attack on the Maersk Line from the NotPetya virus that crippled its terminals worldwide. It was part of a coordinated attack that included the pharmaceutical firm Merck, the Ukrainian government, TNT Express, and a number of other companies. It was rather ironic that it was the Maersk Line that got hit since, up to this point, they had been the industry leader in the digital revolution. It seems that the shipping industry, Maersk included, had not taken the threat seriously enough and had not taken the necessary precautions to upgrade protection against cyber threats. It became clear that shippers, forwarders, terminals, and government agencies were all vulnerable. At the time, Lars Jensen of CyberKeel, a cyber security services provider to the shipping industry, wrote, "This is a situation which is incongruent with the strong drive toward automation and digitization in the industry."[9] Since then, everyone has taken the threat of cyberattacks more seriously. Most notably, a new coalition to address the problem was formed in the Port of New York and New Jersey in early 2020. The idea was to promote information sharing among port users and to conduct port-wide cyber security exercises from time to time. Interestingly enough, there is a whole new field of testing computer systems to find vulnerabilities to hackers. It's called penetration testing.[10] The coalition includes the U.S. Coast Guard, vessel operators, marine terminals, as well as managers from both the energy and financial sectors. The coalition has also set up a Cyber Advisory Committee, drawing on cyber security experts from nearby Rutgers University and the Stevens Institute of Technology. The importance of these initiatives was underscored by Captain Jason Tama, captain of the port, who said, "The nature of the cyber threat demands that different segments of industry, government, and academia share information and work together to enhance cyber preparedness and resilience."[11] Will these new steps completely eliminate all future cyberattacks? That's unlikely, but we can take some comfort in the fact the problem is being seriously addressed by knowledgeable people, rather than being swept aside.

Port Opportunities

I have just touched upon a few of the more obvious risks that may negatively impact ports worldwide in the future, with special reference to Port Newark–Elizabeth. I cannot begin to pretend that I have exhausted the list of future risks. Suffice it to say, that with rapid technological change and increased system integration, it is almost certain that there are risks out there that I have failed to foresee. Rather than continue to speculate about risks, I would like to shift our attention to opportunities that lie ahead in the future. One of the most promising developments to optimize port productivity is the trend toward automation. At first glance, this trend seems almost commonsensical and inevitable. The trend is widespread throughout the American economy. We are all familiar with the demise of toll collectors on our highways. Most of us drive around with an electronic tag mounted on our windshields. Even those without such will get a bill sent to the address linked to their license plate. Elevators in tall buildings no longer have operators; instead, we just push a button. The list goes on and one. According to one study, some 400,000 jobs were lost to automation in U.S. factories from 1999 to 2007.[12]

Why not deploy automation at American ports, including Port Newark–Elizabeth? There are strong arguments to proceed, full speed ahead, to make full use of artificial intelligence and technology. A recent study by Moody's Investor Service lays out a number of reasons for high-volume terminals to adopt automation including a gain in operating efficiency, an improvement in their competitive position, better use of limited land, and lower labor costs. "With the ability to deliver multiple benefits beyond labor savings, we expect adoption of automation to increase both globally and in the United States and Canada," Moody's said in that report of June 24, 2019.[13] A particularly strong advocate for the application of artificial intelligence and technology in support of the whole transportation industry, ports included, is Bradley Jacobs, the CEO of XPO Logistics, headquartered in Greenwich, Connecticut. At a

Traffic Club dinner on February 27, 2019, he said, "Artificial intelligence is the biggest catalyst of change that I see in the world and in our industry. Our industry is basically an information industry. All of the different things that we do have something to do with figuring out where things are, and where they have to go, and what is the best way to get them from here to there. Whether that is by air or by sea or whatever it is, that is what we're trying to do."[14] To be sure, Jacobs is eloquent and persuasive. At that same dinner, he went on to explain that, up to this point, human beings have been doing a satisfactory job in moving the cargo, but, he said, "Human beings with 100 billion brain cells don't hold a candle to what artificial intelligence and do—and will do."[15] Of course, this argument makes sense, but there is a problem. What about the longshore unions that see automation as a mortal threat to their good-paying, blue collar jobs? Jacobs did not seem to be disturbed by the idea of machines taking over these good jobs. He seemed to dismiss that objection by saying, "A lot of the jobs that humans are doing in this industry—let's face it—they're not great jobs. You come home really tired and you don't get paid lots of money. You work in a warehouse, you drive a truck, you work at a cross dock, you work at a port; these are hard jobs and you don't get paid huge amounts of money."[16] Maybe that's all true, but if you depend on that hard job to pay your rent and buy your groceries, you do not want that job to disappear.

To fully understand the clash between terminal operators and longshoremen over the issue of automation, we must pull back and look at the larger picture. According to U.S. Department of Labor statistics, the share of U.S. workers with union membership has fallen from 20 percent in 1985 to 10.5 percent in 2018. What happened? How do we explain this dramatic drop? The customary explanation is that the nature of jobs in the United States has been changing. There has been an increase in jobs in health care, restaurant, and hospitality jobs—sectors without high unionization rates. At the same time, there has been a significant drop in manufacturing jobs—a sector traditionally high in union membership.[17]

Of course, all this is true; but, interestingly enough, there has also been tremendous resistance to unionization among employers in the transportation and supply chain sector. For example, Amazon has successfully resisted unionization within its workforce ever since it was founded back in 1994. The company successfully beat back unionization in Seattle in 2000 and again in Delaware in 2014, both with a wide margin.[18] Meanwhile, FedEx, founded in 1971, with a network of airplanes, shipping hubs, and delivery trucks, has also largely kept unionization at bay. Interestingly enough, FedEx has 280,000

FIGURE 36 Container Terminal Altenwerder (CTA) in Hamburg, Germany, June 5, 2019. This terminal is one of the most modern container terminals in the world. Most of the processes here are automated. Only the unloading of the container from the ship's hold is done by a human being. Everything else—loading the containers onto transport vehicles, transport to the block storage, unloading and later loading the container onto a truck—is automated. (Photo credit: Jose Gribas / Suddeutsche Zeitung Photo / Almay Stock Photo.)

workers, but only 5,000, all airplane pilots, are union members.[19] Without any union opposition, FedEx has been able to install the latest automation at its parcel hub in Memphis, Tennessee. A recent example is the addition of four 260-pound industrial robotic arms. These are equipped with computer vision and artificial intelligence. In other words, they are getting both "eyes" and "brains" that enable them to sense and respond. The new machines help to keep packages moving through the world's largest air freight operation.[20]

What about automating ports in the future? It certainly can be done, and China has often shown the way. For example, there was a successful trial run of a new automated container truck that operated without a driver developed by a group led by Cosco Shipping Ports Limited in May 2020 at the Xiamen Ocean Gate Terminal in China. The truck accurately positioned itself with a cargo crane, which loaded and unloaded containers aboard the vessel *Cosco Shipping Rose*.[21]

When it comes to automating ports in the United States, it is more problematic than we have seen with FedEx and Amazon because of strong longshore unions on both the West Coast and the East. Let us begin with a look at the situation on the West Coast. The union representing dock workers there is the International Longshore and Warehouse Union (LWU), which always viewed automation as an existential threat. At one point, it appeared as if a final solution to the problem had been found. Back in 2008, the ILWU was engaged in negotiations with the Pacific Maritime Association (PMA), the West Coast employer group overseeing longshore labor. At that time, the two parties agreed to a contract that gave terminals the right to automate in return for higher wages and benefits amounting to some $800 million over the period from 2008 to 2022. At the time, union membership approved the contract by a margin of more than 80 percent. According to the agreement, not only did the terminals have the right to automate, but the dockworkers agreed to "not interfere with implementation" of automation. However, as this is written, it appears that the ILWU is now unhappy with the agreement. It appears that the union has been lobbying the California state legislature for a process where there would be a state agency to approve automation projects, one at a time on a case-by-case basis. The prognosis for automation on the West Coast for automation in the future at this point is unclear.[22]

What about port automation on the East Coast? As a general rule, the East Coast has been less aggressive on pushing full automation than the West Coast. The union here is the ILA, which has reluctantly accepted some semiautomation. We find semiautomation at the Global Terminal Container Terminal in Bayonne, New Jersey, and at two terminals in Virginia. These make use of automated stacking cranes (ASCs) in the container yards that operate around the clock with minimal worker involvement. This innovation has been acceptable to the union because the horizontal movement from the dock to the stacks in the yard is done by yard tractors driven by longshore workers. The union here has succeeded in preserving many yard tractor jobs on each shift.[23]

In the distant future, I believe that North American ports will embrace more and more automation if for no other reason than space limitations will force them to make better use of their footprint. However, that's the very long-term view. In the near-term future, things are likely to move slowly. A recent report from Moody's Investor Service warned that the high capital investment required, along with the uncertain productivity gains, makes automation projects risky. The report also addressed labor concerns. There are real political risks to hasty implementation of automation because port authorities are widely seen as job creators. The report warned, "Even if not explicitly stated, there is a limit to the willingness of many port authorities and their parent governments to support automation initiatives that result in meaningful job losses."[24]

We have looked ahead to both the challenges and the opportunities facing the port of the future. Let us now try to do the same for the ships of the future, since ports and ships are inextricably linked in the business of shipping. It is a well-known truism that ships only make money when they sail and offer reliable service. Terminals work best when ships arrive on time, and ships work best when they spend as little time at the dock as possible. In other words, ships and terminals are partners. Both want to be as profitable as possible. In short, shipping is a *business*.[25]

Ship Disruptions

When a ship leaves a port, everyone hopes for a successful voyage. There is a traditional nautical blessing when those ashore wish mariners onboard for "Fair Winds and Following Seas." Yet we all know that the sea holds many perils. Historically, there have been many shipwrecks, yet even now in the twenty-first century, with all of our advances in safety procedures and improvements in navigational technology, accidents continue to happen. Most of these are unlikely to disappear in the future. Let us briefly review some of these hazards, particularly as they affect container ships. There are dozens of such problems, but let us focus on three of the most dramatic—human error on the part of the crew, dangerous cargo resulting in fires, and structural failure of the ship.

We will begin with human error, which can occur due to lack of training, inexperience, fatigue or bad judgment. However, even well-trained crew members can make mistakes if they are physically and mentally exhausted. Overworked ratings and officers are likely to make poor decisions, misread charts, or even fall asleep on the job.[26] Let us consider a particularly compelling example, when the MV *Rena* lost some 900 containers as it ran aground off the coast of New Zealand in October 2011. A subsequent investigation found that the accident was caused by the failure on the part of the master and second officer to follow correct voyage planning that led to the worst maritime disaster in the history of New Zealand.[27]

The record showed that container ship MV *Rena* ran aground in the Bay of Plenty during a voyage from Napier, a coastal city on the country's North Island, to the Port of Tauranga, the largest port in New Zealand, also on the North Island. The ship struck the Astrolabe Reef at 17 knots during the early morning of October 5. The ship was stuck there, and it slowly fell apart. About 200 tons of fuel were spilled, and many cargo containers were lost. Some three

FIGURE 37 Container ship MV *Rena*, approximately fourteen hours after becoming grounded on Astrolabe Reef, off the Port of Tauranga, Bay of Plenty, New Zealand, October 5, 2011. The subsequent investigation found that the accident was caused by human error on the part of the ship's officers. (Photo credit: Brent Stephenson / Alamy Stock Photo.)

months later, the ship broke in two, on the night of January 7, 2012, after being struck by waves of up to twenty feet and heavy winds.[28] Three days later on January 10, 2012, the stern section slipped off the reef. Oil continued to leak into the water.[29]

According to the findings of the accident commission, it seems that the master in charge of planning the duration of the trip simply took the total distance and divided by the average speed of the ship. He failed to take into account unfavorable currents. So the master authorized watch keepers to adjust the course of the ship to avoid such currents. Thus, at one point, the second mate made a decision to deviate from the planned course and shorten the distance, because he was eager to make a meeting early in the morning of October 5 with a pilot boat from Tauranga. That deadline at 3:00 A.M. was necessary because of the port's tidal currents. The course adjustment brought the ship too close to the Astrolabe Reef. At 1:52 A.M., the master returned to the bridge, and assumed control of the ship, unaware of the imminent disaster. According to the report, the master had "received virtually no information on where the ship was, where it was heading, and what dangers to navigation he needed to consider." Then, just a few minutes later, the ship struck the reef at 2:14 A.M.[30]

Sometimes things go wrong that are no fault of the crew onboard the ship. Indeed, there are cases of explosion and fire, where it is nearly impossible to find out whom to blame because so much of the evidence has been destroyed. For example, consider the case of the container ship *Hanjin Pennsylvania*, a German-owned ship under South Korean management. This brand-new ship was on its way from Singapore to Hamburg, Germany. Then, on November 11, 2002, while sailing about ninety miles south of the Sri Lankan coast, there was a report of an explosion. Two men died, and the rest of the crew had to abandon the ship later in the day, after being unable to put out the resulting fire. The fire continued to burn for six weeks. Finally, the salvors put out the fire, and towed the ship to Singapore in January 2003. About half of the containers were destroyed in the fire.[31]

In the subsequent investigation, it turned out that the ship had been carrying a large consignment of fireworks, stored in some fifty-seven containers. During the fire, there was a spectacular display. One might think: okay, here's the problem—the ship was carrying fireworks. It turns out that it was not that simple. Research shows that most of the fireworks used in Europe are imported from China. Consequently, maritime transport of fireworks is routine, provided that the manifest lists them as hazardous goods. Some 8,000 twenty-foot equivalent containers of fireworks arrive in Europe every year.[32]

Fire investigators were never able to pinpoint the cause of the fire, which was one of the largest marine insurance claims of 2002, even after having removed all the containers from the ship. Rather surprisingly, it seems unlikely that the fire started with the fireworks. There is a strong suspicion that the fire may have actually started with a consignment of calcium hypochlorite stored below decks. One expert said, "Although dangerous goods aided the spread of the fire, it may be that the accident was not caused by such cargoes. Also the damage to the ship and cargo was so great that we may never be able to determine what the root cause of the accident was." At first, the ship was declared a total loss and sold for scrap.[33] However, the hull was purchased by a British company that contracted with a Chinese firm to have the ship rebuilt. It is now sailing as the *Nova Bellatrix*.[34]

Finally, in our brief discussion of container ship disruptions, let us take up the story of the loss of the Bahamian flagged, post-Panamax container ship MOL *Comfort*, said to be the worst shipping in history. It was in June 2013, when the ship, chartered by Mitsui OSK Lines of Japan, was crossing the Indian Ocean, going from Singapore to Jeddah in Saudi Arabia. It was on June 17, 2013, when the *Comfort* encountered extremely bad weather about 200 miles off the cost of Yemen. Suddenly disaster struck, and the ship broke in half. Initially, the two parts stayed afloat, with most of the cargo safely preserved. A plan was developed for Smit International, a salvage company, to save the two halves of the container ship by using tugboats to bring them to shore, thus saving all of the cargo. The plan quickly unraveled. On June 27, some ten days later, the stern section sank to a depth of 4,000 meters, taking with it 1,700 shipping containers and 1,500 tons of fuel. Then, on July 6, a fire broke out in the bow section, which later sank on July 11 to a depth of 3,000 meters. Insurance claims after the disaster amounted to somewhere between $300 and $400 million.[35]

So what happened? How to account for this unexpected structural failure? After the disaster, ClassNK, a Japanese organization that maintains technical standards for the construction and operation of ships, conducted a detailed investigation. They conducted a study taking into account the stresses that the heavy seas placed on the ship's structure. It is well known in the trade that, in heavy weather, the ship will go from a "hogging" (the hull curves upward in the middle) to a "sinking" (curves downward) condition, over and over again. Of course, hogging and sagging describe only the fore and aft stresses on the ship. We know that lateral forces are also in play, compounding the forces on the hull structure.[36]

The classification society concluded that the hull structure conformed to its technical requirements, and the ship should have survived the heavy weather. In theory, the MOL *Comfort* had more than enough hull girder strength, yet still the ship broke up. The final lengthy technical report struggled with identifying the cause of the crack failure. It dealt with many uncertainties. One strong possibility is that there was a difference between the declared weights and the actual weights of some of the containers, which may have contributed to the fracture.[37]

Ship Opportunities

We have just briefly examined the dangers posed by human error, shipboard fires, and structural failures of ships. It seems that risk is deeply embedded in the shipping business. As we look to the future, we hope to see these hazards minimized. At the same time, we know that danger will never be completely eliminated. Let us now shift gears, and take a look at one of the most promising opportunities that lies ahead in the future. That would be automation. Will we ever see totally automated container ships? Perhaps, but it is a long way off. Nonetheless, there have been some promising first steps.

As we look back at maritime history, there have been many changes. We evolved from sail to steam and then onto diesel. As this is written, it appears that we are in the midst of yet another transition—this time to automation. It is a period that is both fascinating and challenging. For some applications, the benefits of automation are so compelling that there has been a big push to adopt it and to proceed. For others, there are simply too many drawbacks. Perhaps the most obvious place for maritime automation is within minesweepers. These are small warships tasked with the removing of explosive naval mines by using various technologies to detonate the mines and to clear the path for other ships. Clearly, this is a dangerous activity, and it would be desirable to remove the crew altogether. In early 2020, the Royal Navy adopted an unmanned minesweeper with the name of ARCIMS, which stands for Atlas Remote Combined Influence Minesweeping System. It was developed by Atlas Elektronic and designed to set off underwater mines without exposing any crew to danger. It works by towing three smaller boats, each with acoustic, magnetic, and electronic devices that cause the enemy mines to explode at a safe distance.[38] By way of stark contrast, it is unlikely that passenger vessels (even for short trips like the Staten Island Ferry) will ever become automated. On these vessels, you need crew members on hand for routine tasks such as repeatedly tying and untying lines, boarding passengers, and assisting customers, not to mention other duties such as making minor repairs or refilling the toilet paper dispensers in the restrooms. More importantly you want to have someone available to stop suicides or to throw those who fall overboard a life ring.[39]

A review of the literature indicates that maritime autonomous surface ship technology is a hot topic, and it is being rapidly developed around the world. Although it has been more slowly embraced in the United States than elsewhere, the United States is beginning to catch up. There are promising developments in the use of smaller unmanned vessels, especially in applications that are dangerous or tiring, such as oil spill response and firefighting. Even here, automation does not necessarily mean totally eliminating the crew. It may simply mean reducing the size of the crew; hence, consequently, reducing the exposure to risk. To be sure, this means fewer jobs. However, unlike what we have seen with U.S. ports, there are really no strong maritime unions to oppose automation. It would seem that the main problem to the widespread adoption of autonomous surface ships is the matter of law. Existing U.S. and international regulations were drafted without any anticipation of automation. So we are faced with questions such as who is in charge or who is responsible. In the years to come, the International Maritime

FIGURE 38 ASV *Global*, an unmanned autonomous surface vessel being demonstrated at the Oceanology International 2018 Exhibition in London's Royal Docks, March 14, 2018. Advocates believe that there is a future for such small autonomous vessels to perform various tasks, but there are numerous challenges to developing large autonomous vessels. (Photo credit: AJBC_1 / Alamy Stock Photo.)

Organization will have to address these questions and come up with new regulations.[40]

Taking advantage of the new opportunities opened up with autonomous surface ships are a number of relatively new companies. One of them is ASV *Global*, whose name is derived from the term "autonomous surface vehicles." For better or worse, it seems that the use of such acronyms is ubiquitous and unavoidable in the shipping business. In any event, the company develops, builds, and operates these ASVs with facilities in Portchester in the United Kingdom and in Broussard, Louisiana, in the United States. The company has taken justifiable pride in being a cutting-edge leader in this innovative technology. For example, in August 2015, ASV *Global* provided the very first autonomous surface vehicle to the operator, TerraSond, to perform bathymetry for updating U.S. nautical charts. This important study of the underwater depth of the ocean floor took place in the Alaskan arctic. The 49-foot ASV surveyed alongside the operator's mother vessel, effectively doubling the collection of data. Perhaps even more important, the ASV was able to survey by itself areas that were either too shallow or too dangerous for the larger ship.

Tom Newman, president of TerraSond, said, "This is a force-multiplier for data acquisition. Operated in a semi-autonomous mode, unmanned but supervised, one person can replace the three person crew it would normally take to operate a survey launch; it is definitely the future of seafloor mapping."[41]

From the above example, it is clear that there is a future for relatively small autonomous vessels to perform various tasks. But can they carry cargo? Well, it turns out that on May 7, 2019, customs officials in Ostend, Belgium, received a shipment of one box of oysters from Essex in the United Kingdom. It has been called "the world's first unmanned commercial shipping operation." The boat, without a crew, was carefully supervised by four people in a command center in Tollersbury, Essex, home of Hushcraft, the company that designed and developed the vessel. The vessel was constantly feeding information to the command center by means of onboard cameras and microphones. This historic voyage was monitored by the coastguards of both the United Kingdom and Belgium. Ben Simpson, managing director of Hushcraft said, "The benefits are many. You can send them around the world to do different jobs for a significantly reduced cost. Then you do not have to have a galley; you do not have to have toilets. You can utilize space."[42]

Granted, one box of oysters hardly compares with a giant container ship with thousands of boxes. Nonetheless, this was a key historical moment. But before we get ahead of ourselves, there are caveats. Lawrence Brennan, a retired U.S. Navy captain and adjunct professor of maritime law at Fordham University, agrees that having no crew means no risk to human life and no need to recruit and pay a highly trained crew. At the same time, he goes on to warn that a breakdown in communication between the base and the vessel would create a helpless ghost ship. He went on to point out another looming problem—the law. Professor Brennan said, "The legal regime is decades, if not a century-and-a-half out of date. As unmanned ships were never contemplated until recently, legislation says manning is essential for having a ship that is seaworthy, classified, and authorized to operate in national waters and on the high seas."[43]

For all the excitement, complete automation of cargo ships in general, and container ships in particular, is far off in the future. It may work for small ships with small cargoes going short distances, but there are many problems for implementing it with transoceanic container ships. For the foreseeable future, we will need onboard decision makers, repair experts, and security guards in order to operate efficiently and safely. Take the matter of repair. A ship is a box with hundreds of machines, most of which are in need of routine maintenance. For example, a regular fuel filter has between 200 and 500 hours of useful life, and a clogged filter can shut down the whole show. Someone has to be on hand to change oil or add oil as needed. Then there is the matter of lighting. A regular old-fashioned filament light bulb in maritime use has a 1,000 hour life span, in other words about two months. Of course, LED lights are good for about two years, but even these must be changed as needed. Without some sort of maintenance crew with wrench turners, oil adders, light bulb changers, and computer rebooters, we might find our expensive ship dead in the water.[44]

To be sure, the matter of onboard repairs gives us serious pause about ship automation; however, at the same time, there have been some promising developments in things like camera technology, sensors, artificial intelligence, and satellite technology. One such recent development stands out. Pomare, a marine research outfit, has teamed up with IBM to develop an artificial intelligence captain (AI Captain) to take the helm of an experimental ship, *Mayflower*. The three-hulled trimaran was built in Gdansk, Poland. It is an exciting experiment. The AI Captain is expected to use cameras, artificial intelligence, and advanced computing systems to sense, think, and make decisions, while safely navigating around ships, buoys, floating containers, and other hazards. Rob Hugh, vice president for IBM Edge Computing, said, "The Mayflower needs to sense its environment, make smart decisions about its situation and then act on these insights in the minimum amount of time—even in the presence of intermittent connectivity, and all the while keeping data secure from cyber threats."[45]

The *Mayflower* experiment seems to hold out the promise of cost savings for not having to pay for officers and crew. But there is a long way to go to implement autonomous systems on a widespread basis for commercial cargo ships. A leading skeptic is Commander David Dubay, USCG, who has written convincingly that there are simply too many problems. He points out risks posed by hackers, terrorists, and criminals. These risks are compounded if they are no humans on the scene to respond. He also points out the difficulties in developing a foolproof command and control system: "Weather, wind, waves, fog, obstructions, marine mammals, salt water, birds, other ships, sounds, and almost anything else imaginable is encountered out on the open ocean. An autonomous ship will require incredibly complex technology to withstand the chaos of the ocean environment and enable a ship to respond to any incident or emergency."[46]

Commander Dubay is eloquent and persuasive in his argument that autonomous shipping is a long way off and may never happen. From my research, I believe that his point of view is representative of the contemporary

American maritime community. Yet, at the same time, eight of the other leading maritime nations have come together to encourage the development of what they have been calling maritime autonomous surface ships. Alas, this means that they have come up with another acronym, MASS. To go along with this, they have also come up with the sister acronym MASSPorts, which is the idea of getting ports ready for autonomous shipping. The movement has representatives from China, Denmark, Finland, Japan, the Netherlands, Norway, the Republic of Korea, and Singapore. Noticeably absent is the United States. These eight countries have been joined also by representatives from International Maritime Organization, the International Association of Marine Aids to Navigation and Lighthouse Authorities, and the International Association of Ports and Harbors. The whole idea is to try to get ready for the port of the future. Commenting on this development, Rene de Vries, harbor master of the Port of Rotterdam Authority, said, "We are aware of the fact that conventional shipping will remain a reality in our port, but we are also convinced that smart ships and even autonomous ships will be visiting Rotterdam as well. Smart ships need smart ports. Innovative co-operation between ports, the industry, and shipping is key."[47] With all this in mind, I predict that, in the proverbial long run, autonomous shipping will prevail, though I may not live long enough to see it happen.

Acknowledgments

The origins of this book go back a long way. For many years, I had been teaching courses in maritime history and culture at Rutgers, often combining classroom instruction with various fieldtrips including visits to Port Newark–Elizabeth. As I spent more time at the port with all of its complexity, I began to think of it as the possible subject of a book. As I started the project, I consulted the book *The Airport: Planes, People, Triumphs, and Disasters at John F. Kennedy International* by James Kaplan, published by William Morrow. Kaplan approached the topic inspired by his curiosity about how it all worked. I began to think that, if Kaplan can do that for an airport, perhaps I could do the same thing for a seaport.

The idea really began to come together for me in 2006, when a number of books were published about the rise of containerization. These included *The Box That Changed the World: Fifty Years of Container Shipping—An Illustrated History* by Arthur Donovan and Joseph Bonney, published by the *Journal of Commerce*; *The Box: How the Shipping Container Made the World Smaller and the World Economy Bigger* by Marc Levinson, published by Princeton University Press; and *Box Boats: How Container Ships Changed the World* by Brian J. Cudahy, published by Fordham University Press. I was inspired by these three books, and I am indebted to them for underscoring the global importance of containerization.

As time went on, there were other books that fueled my interest in the topic. There was *Getting the Goods: Ports, Labor, and the Logistics Revolution* by Edna Bonacich and Jake B. Wilson, published by Cornell University Press.

These coauthors wrote about the twin ports of Los Angeles and Long Beach in Southern California, much as I was planning to write about the twin ports of Newark and Elizabeth in northern New Jersey. Next there was *Ninety Percent of Everything: Inside Shipping, the Invisible Industry That Puts Clothes on Your Back, Gas in Your Car, and Food on Your Plate* by Rose George, published by Henry Holt. In this book, the author focused more on the container ships themselves rather than the ports, and she shared her experience riding along and observing the movements of the container ship *Maersk Kendal*. Finally, there was *Door to Door: The Magnificent, Maddening, Mysterious World of Transportation* by Edward Humes, published by HarperCollins. Humes deals with ships and ports in the larger context of the entire world of transportation including trucks, trains, and airplanes.

In writing the early history of Port Newark, I depended heavily on accounts found in the *New York Times*, which has so often been called "the newspaper of record." It was particularly helpful, for example, in telling the story of dramatic development of the port in the 1920s. Going forward in time, I relied on the *Journal of Commerce*, a specialized biweekly magazine with a focus on international logistics and shipping. It was especially helpful useful in dealing with contemporary topics such as the trade war with China and the ongoing problem of cyberattacks.

A word of thanks is in order concerning the staff at Rutgers University Press. My wholehearted thanks to director Marlie Wasserman, who gave me early encouragement and support. Thanks also to her successor Micah Kleit and executive editor Peter Mickulas, who gave me the green light to proceed. I am indebted to Michael Siegel, staff cartographer of the geography department of Rutgers University for drawing the map of Port Newark that serves as the frontispiece for the book. Thanks to Dustin Braden, shipper engagement manager at the *Journal of Commerce*, for providing a selection of historic photographs of the SS *Ideal X*, the first commercially successful container ship. Thanks also to maritime photographer David Rider for a selection of photographs documenting contemporary harbor operations. Special thanks to New York City freelance photographer Leo Sorel for the formal portrait of the Rev. David Rider taken on the roof of the International Seafarers' Center at Port Newark.

Writing a book takes hard work and self-discipline, something I have in short supply. It can be difficult and lonely. For me, the best part of this project was getting out of my study and going out to the harbor—meeting people, making friends, and conducting interviews. It is the interviews that make this story come alive, and I am especially grateful to all those who took the time to meet with me and explain things. Of course, to set the stage for the rise of containerization, I had to do some background research on what came before, namely break-bulk shipping, moving the freight in separate containers. Most of this was historical research, but I had the good fortune to interview retired marine carpenter Frank Greco who worked at Port Newark from 1951 to 1953. Greco gave me a good firsthand explanation of how things were done before containerization.

By the time that I came along, it was too late to interview Malcom McLean, who died in 2001. However, I had the good fortune to speak with two of his key lieutenants. First, there was naval architect Charles R. Cushing, who patiently explained to me Malcom McLean's thrifty pattern of adopting old ships for containerization rather than buying new ones. Next, I interviewed civil engineer Ronald Katims, who told me the story of how he went about designing Elizabeth Marine Terminal as a container port from scratch with no precedents. He then went on to detail McLean's rapid expansion into Puerto Rico, Oakland, Long Beach, Alaska, and even Vietnam.

In piecing together the history and geography of New York Harbor as well as the approach to Port Newark and Port Elizabeth, I received much good advice from the staff of the National Lighthouse Museum in Staten Island. Special thanks to Linda Dianto, executive director of the museum, who answered many of my questions.

I have a great debt to several long-term members of the U.S. Coast Guard, who have played various roles in getting ships safely from the ocean to the port. These include Torrey Jacobsen, who served as commanding officer of the *Katherine Walker*, a coastal buoy tender. Jacobsen took the time to fill me in on the rich details of buoy maintenance. In addition, I received a great deal of help, in a series of interviews, from Jason Wiley, who served as officer in charge of the Aids to Navigation Team (ANT), located at the Military Ocean Terminal in Bayonne, New Jersey. I would also like to thank Greg Hitchen, director of the Vessel Traffic Service at Coast Guard Sector New York. I had the opportunity to visit Hitchen at his office located at historic Fort Wadsworth on Staten Island. Hitchen opened the door for me to visit the highly secure area, where watch standers coordinate vessel movements in the harbor. There I was able to interview Carrieann Dixon, watch supervisor, who patiently explained some of the typical problems on a given day.

When it came time to explain the role of the maritime pilots, I received an extraordinary amount of help from two of the practitioners. First was Sandy Hook pilot John Oldmixon, an experienced navigational expert in getting ships in from the open ocean to the relatively safer waters of New York

Harbor. Next was senior harbor pilot Simon Zorovich, who patiently explained how he goes about guiding ships through the congested harbor and getting them safely docked. Both of these gentlemen had a knack for explaining something complicated to someone like me who had little prior knowledge of the topic. I am indebted to them both. Meanwhile, I had long been fascinated with the role of tugboats, those small and powerful watercraft that assist docking pilots in maneuvering large ships into a dock. To fill me in on tugboat operations, I turned to tugboat officer Daniel Zorovich, the son of our harbor pilot. Like his father, Daniel was kind, patient, and thorough in explaining the life and work of tugboat operators.

I have spent many years studying and writing about facilities built and operated by the Port Authority of New York and New Jersey. My book *Twin Towers: The Life of New York City's World Trade Center* was published in 1999. This was followed in 2011 with the publication of *Crossing under the Hudson: The Story of the Lincoln and Holland Tunnels*. In both cases, I found officials at the Port Authority to be responsive and cooperative—returning my phone calls and setting up interviews. It was much the same in conducting research for this book on Port Newark. Here, I am especially indebted to Bethann Rooney, the deputy director of the Port Department.

Many different types of cargo are handled at Port Newark and Port Elizabeth. These include such varied items as automobiles, orange juice, and Belgian blocks—all of which have interesting background stories. However, in this book, I have tried to keep the focus on containerization. Thus I was especially fortunate to have access to the APM in Port Elizabeth, thanks to Giovanni Antonuccio, general manager of client services. He patiently explained everything including how the dockside cranes load and unload the ships simultaneously, as well as how the yard hustler trucks move containers to and from the storage yard.

Closely related to the container ports are the specialized trucking companies that bring the containers in and out of the ports. To better understand their operation, I turned to Dick Jones, who serves as the executive director of the Association of Bi-State Motor Carriers. I owe a particular debt to Jones, who explained in detail the work that these truckers perform and the difficulties that they encounter along the way. Also very helpful in understand all this was Philip Gigante of BBT Logistics in Newark, New Jersey.

The Seamen's Church Institute (SCI) operates a center at the heart of Port Newark. It serves an important role in providing programs and services for the crews of visiting ships. I owe a great deal of thanks to the staff of SCI. It was here that I launched my study of the port with their help. They welcomed me, showed me around, and introduced me to many of the key players. The center became my home away from home, while I was conducting interviews around the port. I am greatly indebted to the Rev. Jean R. Smith; the Rev. David M. Rider; the Rev. Megan Sanders; the Rev. James Kollin; Cora DiDomenico, M.Div.; Michelle McWilliams, MSW; Timothy Wong, director of the International Seafarers' Center, Port Newark, New Jersey; and Douglas Stevenson, Esq.

When it came time to speculate about the future, I gave a good deal of thought to the wave of automation now sweeping through port management and ship development. On the positive side, there are potential savings and efficiencies. Having said that, I am indebted to Captain Robert Quigley for pointing out the practical downsides of automation as well as Lawrence Brennan, Esq., for pointing out the legal downsides.

As I look back on the writing of this book, I realize how much I have relied on interviews with key figures. I am reminded that I owe a debt of gratitude to the late Kenneth Goldstein, who encouraged me, along with his other graduate students at the University of Pennsylvania, to overcome our shyness and get out there and conduct interviews. Another early guide who understood my interest in the Port Authority of New York and New Jersey was the late Jameson W. Doig of Princeton University. He gifted me with encouragement, suggestions, and contacts. In the scholarly community, I have received supportive advice on the writing process from James T. Fisher, Nikolai Burlakoff, Ben Sifuentes, Louis Masur, Jorge Schement, and Michael Aaron Rockland.

My thanks to Rutgers, the State University of New Jersey, for two sabbatical leaves from teaching duties during the fall semesters of 2012 and 2019. These were two wonderful periods of uninterrupted time allowed me to focus on the research for this book.

Special thanks to my wife, Rowena Cosico Gillespie, whose enthusiastic support for this book was key. You were my motivation and inspiration. Thank you for everything.

Notes

Chapter 1   Early Historical Background

1. John T. Cunningham, *Newark* (Newark: Newark Historical Society, 1966), 103–104; "History of Newark."
2. Cunningham, *Newark*, 114–124.
3. Cunningham, *Newark*, 172–173.
4. Cunningham, *Newark*, 246.
5. Kenneth C. Davis, *Don't Know Much about Geography* (New York: William Morrow, 1992), 298.
6. "Make Newark Bay a Big Port," *New York Times*, June 27, 1915; Cunningham, *Newark*, 246–247.
7. "Make Newark Bay a Big Port," *New York Times*, June 27, 1915.
8. "Make Newark Bay a Big Port," *New York Times*, June 27, 1915.
9. U.S. War Department, "Annual Report of the Secretary of War 1919" (Washington, DC: Government Printing Office, 1920), vol. 1, pt. 4, 4136.
10. George F. Shephard, "Uncle Sam: Wholesale Warehouseman: Big Army Base in Brooklyn," *New York Times*, December 7, 1919.
11. John Pike, "Newark Bay Shipyard," *Global Security*, http://www.globalsecurity.org/military/facilities/newark-bay.htm.
12. Pike, "Newark Bay Shipyard."
13. Submarine Boat Corporation, "Launching Program for SS Agawam" (May 30, 1918), Records and Manuscript Collections, Rensselaer Polytechnic Institute, http://archon.server.rpi.edu/archon/?p=digitallibrary/digitalcontent&id=32.
14. Submarine Boat Corporation, "Launching Program for SS Agawam."
15. Submarine Boat Corporation, "Launching Program for SS Agawam."
16. "First Fabricated Ship Accepted," *Shipping* 5 (October 26, 1918).
17. Julius Henry Cohen, *They Builded Better Than They Knew* (New York, Julian Messner, 1946), 275.
18. Cohen, *They Builded Better*.

19 Cohen, *They Builded Better*.
20 Jameson W. Doig, *Empire on the Hudson* (New York: Columbia University Press, 2001), 28.
21 Cohen, *They Builded Better*, 276–279.
22 Doig, *Empire on the Hudson*, 45–46.
23 Doig, *Empire on the Hudson*, 1–110.
24 "Film Shows Need of Improving Port," *New York Times*, July 29, 1921.
25 James T. Fisher, *On the Irish Waterfront* (Ithaca, NY: Cornell University Press, 2009), 55–62.
26 Doig, *Empire on the Hudson*, 143–177.
27 "The City of Newark to Expend $1,250,000 in the Development of Newark Bay as a Seaport," *New York Times*, July 31, 1921.
28 "Predicts Active Newark Market," *New York Times*, January 1, 1922.
29 "Deal on for Big Plant," *New York Times*, January 13, 1926.
30 "Port Newark's Statistics," *New York Times*, January 11, 1927.
31 "Government Lets Port Newark Base," *New York Times*, December 2, 1926.
32 "Say Lighterage Fees Would Penalize Port," *New York Times*, June 3, 1929.
33 "Port Newark Sees Progress in 1928," *New York Times*, December 26, 1927.
34 C. J. Fagg, "Letter to the Editor," *New York Times*, December 23, 1929.
35 Cunningham, *Newark*, 280.
36 "Newark to Lose Its Ocean Fleet," *New York Times*, June 28, 1931.
37 "Port Work Lags for Lack of Funds," *New York Times*, January 25, 1931.
38 "Newark Approves Lease of Big Tract," *New York Times*, December 31, 1936.
39 "Says Free Lighterage Is Blow to Newark," *New York Times*, October 23, 1930.
40 "Tonnage Loss Seen in Lighterage Case," *New York Times*, May 8, 1940.
41 "Free Lighterage Called a 'Racket,'" *New York Times*, June 26, 1940.
42 Doig, *Empire on the Hudson*, 187–188; Cunningham, *Newark*, 285.
43 "Navy Purchasing Newark Shipyard," *New York Times*, January 25, 1942.
44 Cunningham, *Newark*, 294.
45 *Jane's Fighting Ships of World War II* (New York: Crescent Books, 1989), 280; and Joel Baglole, "Naval Destroyers—Protecting the Fleet," http://usmilitary.about.com/od/navyweapons/a/navydestroyer.htm?p=1.
46 John R. Ward, "The Little Ships That Could" (Destroyer Escort Sailors Association), http://www.desausa.org/de_program_history.htm.
47 Destroyer History Foundation, "Cannon Class," http://destroyerhistory.org/de/cannonclass/index.asp?r=200&pid=230.
48 Commander X, *The Philadelphia Experiment Chronicles* (Wilmington, DE: Abelard Productions, 1994), 5–6.
49 Naval History and Heritage Command, "The Philadelphia Experiment," http://www.history.navy.mil/faqs/faq21-2.htm.
50 Internet Movie Database, "The Philadelphia Experiment (1984)," http://www.imdb.com/title/tt0087910/synopsis?ref_=tt_stry_pl.
51 Internet Movie Database, "The Philadelphia Experiment (2012)," http://www.imdb.com/title/tt2039399/.

Chapter 2   The Post-World War II Era

1 John T. Cunningham, *Newark* (Newark: Newark Historical Society, 1966), 278–301.

2. Jameson W. Doig, *Empire on the Hudson* (New York: Columbia University Press, 2001), 258–265; Joan Cook, "Harland Bartholomew, 100, Dean of City Planners," *New York Times*, December 7, 1989.
3. Doig, *Empire on the Hudson*, 265–267.
4. Doig, *Empire on the Hudson*, 271–273.
5. "New Plan Offered on Newark Air Site," *New York Times*, September 10, 1946.
6. Doig, *Empire on the Hudson*, 278.
7. "Lease of Airport Backed in Newark," *New York Times*, October 5, 1947.
8. "Battle Dims Hope on Newark Port," *New York Times*, October 16, 1947.
9. Doig, *Empire on the Hudson*, 281.
10. "45 Acres, Several Warehouses Added to Port Newark for Restoration to Public," *New York Times*, July 11, 1948.
11. "Navy Leases Land to Port Authority," *New York Times*, September 29, 1948.
12. "New York Authority Plans $7,000,000 Issue to Finance Development of Newark Seaport," *New York Times*, November 18, 1948.
13. "Big Gains Reported by Port Authority," *New York Times*, January 16, 1949.
14. "Port Newark's Progress," *New York Times*, May 24, 1950.
15. "Cargo Terminals in Newark Opened," *New York Times*, May 25, 1950.
16. Brian Catchpole, *The Korean War* (New York: Carroll & Graf, 2000), 1–17; Bruce Cumings, *The Korean War: A History* (New York: Modern Library, 2010), 5–23.
17. Frederick C. Lane, *Ships for Victory: A History of Shipbuilding under the U.S. Maritime Commission in World War II* (Baltimore: Johns Hopkins University Press, 1951), 42–45, 72–77.
18. Peter Elphick, *Liberty: The Ships That Won the War* (Annapolis, MD: Naval Institute Press, 2001), 37.
19. Lane, *Ships for Victory*, 574–587.
20. Lane, *Ships for Victory*, 27–32.
21. Marc Levinson, *The Box: How the Shipping Container Made the World Smaller and the World Economy Bigger* (Princeton, NJ: Princeton University Press, 2006), 16–21.
22. Frank Greco, interview with the author, East Brunswick, NJ, September 7, 2013.
23. Charles Zerners, "Big Port Terminal Near Completion," *New York Times*, January 31, 1954.
24. "Big Marine Depot Opens in Newark," *New York Times*, March 26, 1954.
25. "Big Ship Terminal Slated in Jersey," *New York Times*, December 1, 1955.
26. George Cable Wright, "Huge New Port Elizabeth to Be Built in Marshland," *New York Times*, December 2, 1955.
27. James T. Fisher, *On the Irish Waterfront* (Ithaca, NY: Cornell University Press, 2009), 250–276.

Chapter 3 The Invention of Containerization

1. Arthur Donovan and Joseph Bonney, *The Box That Changed the World* (East Windsor, NJ: Commonwealth Business Media, 2006), 51–52.
2. Marc Levinson, *The Box: How the Shipping Container Made the World Smaller and the World Economy Bigger* (Princeton, NJ: Princeton University Press, 2006), 47–48.
3. "Tankers to Carry 2-Way Pay Loads," *New York Times*, April 27, 1956.
4. Given its importance, the story of McLean has been well documented. My account is based on a number of sources beginning with my own journal article "History of

the Containership," *Maritime Reporter and Engineering News*, October 1999, anniv. ed., 1A–16A. Jameson W. Doig has given the most authoritative account in his classic *Empire on the Hudson* (New York: Columbia University Press, 2001), 374–376. Thanks to Doig for calling attention to the correct spelling of Malcom. I am also indebted to Anthony J. Mayo and Nitin Nohria of the Harvard Business School for their essay "The Truck Driver Who Reinvented Shipping," *Research and Ideas*, October 3, 2005, 1–7, http://hbswk.hbs.edu/item/5026.html.

5. Mayo and Nohria, "Truck Driver," 2.
6. Robert Mottley, "The Early Years," *American Shipper*, May 1996, 28.
7. Mottley, "Early Years," 30.
8. Mottley, "Early Years," 30.
9. Margalit Fox, "Keith Tantlinger, Builder of Cargo Containers, Dies at 92," *New York Times*, September 6, 2011.
10. Charles R. Cushing, interview with the author, Spring Lake, NJ, June 2, 2018.
11. Mayo and Nohria, "Truck Driver," 5.
12. Megan Scully, "Maritime Matters," *Sea Power Magazine*, June 2018, 9.
13. Ronald Katims, interview with the author, Edison, NJ, May 24, 2018.
14. Brian J. Cudahy, *Box Boats: How Container Ships Changed the World* (New York: Fordham University Press, 2006), 27.
15. Cudahy, *Box Boats*, 28.
16. Alex Roland, W. Jeffrey Bolster, and Alexander Keyssar, *The Way of the Ship: America's Maritime History Reenvisioned, 1600–2000* (Hoboken, NJ: John Wiley, 2008), 346–347.
17. Cushing interview.
18. Cudahy, *Box Boats*, 32.
19. "Ten New Vessels to Haul Trailers," *New York Times*, July 11, 1957.
20. Phillip L. Zweig, *Wriston: Walter Wriston, Citibank, and the Rise and Fall of American Financial Supremacy* (New York: Crown, 1995), 78.
21. Zweig, *Wriston*, 81.
22. Katims interview.
23. Arthur H. Richter, "Trailership Run to Houston Opens," *New York Times*, October 4, 1957.
24. Katims interview.
25. Katims interview.
26. Katims interview.
27. Lowndes County Historical Society Museum, "How Did Valdosta Become the Azalea City?," http://valdostamuseum.com/exhibitions/online-exhibits-2/how-did-valdosta-become-the-azalea-city/.
28. Michael T. Pasquier, "Jean-Baptiste Le Moyne, Sieur de Bienville," in *Encyclopedia of Louisiana*, ed. David Johnson (Louisiana Endowment for the Humanities, 2010–), http://www.knowlouisiana.org/entry/jean-baptiste-le-moyne-sieur-de-bienville-2.
29. Bud Feurer, "Raphael Semmes & the Battle Off Cherbourg," *Sea Classics*, August 2017, 52–64.
30. American Battlefield Trust, "P.G.T. Beauregard," https://www.battlefields.org/learn/biographies/p-g-t-beauregard.
31. Poem Hunter, "John Boyle O'Reilly," https://www.poemhunter.com/poem/from-that-fair-land-and-drear-land-in-the-south/.
32. Donovan and Bonney, *Box that Changed the World*, 68.

33 Donovan and Bonney, *Box that Changed the World*, 61–62.
34 Donovan and Bonney, *Box that Changed the World*, 61–63.
35 Jacques Nevard, "Trailership Cutting Pier Loading Time," *New York Times*, November 23, 1958.
36 Nevard, "Trailership Cutting Pier Loading Time."
37 Werner Bamberger, "Industry Is Exhibiting Caution on Containerization of Fleet," *New York Times*, December 4, 1960.
38 "U.S. Body Enters Container Field," *New York Times*, April 28, 1961.
39 Bernard Stengren, "Containers Display Here Shows Field's Advances and Problems," *New York Times*, September 10, 1961.
40 Cudahy, *Box Boats*, 77–79.
41 Edward A. Morrow, "Sea-Land Will Get 4 Vessels for Intercoastal Trailer Runs," *New York Times*, July 8, 1962.

Chapter 4   The Rapid Growth of Containerization

1   Charles R. Cushing, interview, Spring Lake, NJ, June 2, 2018.
2   Cushing interview.
3   "Lyle King, Leader in Port Authority," *New York Times*, December 11, 1973.
4   Jameson W. Doig, *Empire on the Hudson* (New York: Columbia University Press, 2001), 375.
5   Marc Levinson, *The Box: How the Shipping Container Made the World Smaller and the World Economy Bigger* (Princeton, NJ: Princeton University Press, 2006), 91–92.
6   Joseph O. Haff, "Huge Marine Job Begun in Jersey," *New York Times*, July 16, 1958.
7   "Jersey Piers Set for New Channel, *New York Times*, July 15, 1960.
8   George Cable Wright, "Authority Spurs Port Elizabeth," *New York Times*, March 23, 1962.
9   Ron Katims, interview, Edison, NJ, June 14, 2018.
10 Katims interview.
11 Katims interview.
12 Jean-Paul Rodrigue and Mark Booth, "Grounded and Chassis Container Terminal Operations" (Port Technology International, n.d.), https://people.hofstra.edu/jean-paul_rodrigue/downloads/Booth%20&%20Rodrigue_PT57_V5.pdf.
13 Katims interview.
14 Katims interview.
15 "Ultra Modern Port Operation," *World Port and Marine News*, May 1965, 34.
16 Katims interview.
17 Katims interview.
18 Katims interview.
19 Ron Katims, telephone interview, July 9, 2018.
20 "Ultra Modern Port Operation," 35.
21 Katims telephone interview.
22 Ron Katims, "Malcom Purcell McLean," undated and unpublished manuscript in possession of the author.
23 Katims telephone interview.
24 Katims, "Malcom Purcell McLean."
25 Ron Katims, "Alaska," undated and unpublished manuscript in possession of the author.

26 USGS, "The Great M9.2 Alaska Earthquake and Tsunami of March 27, 1964," https://earthquake.usgs.gov/earthquakes/events/alaska1964/.
27 Katims, "Alaska."
28 Katims telephone interview.
29 David P. Yens and John P. Clement III, "Port Construction in Vietnam," *Military Engineer* 59, no. 387 (January–February 1967): 20–24.
30 "DeLong Piers, 1967 U.S. Army Research and Development Progress Report 9," YouTube, https://www.youtube.com/watch?v=GG3B_vM49TY.
31 "DeLong Piers."
32 "Delong Pier Construction at Cam Ranh Bay, 1966 U.S. Army, Vietnam War," YouTube, https://www.youtube.com/watch?v=313AGcHcHkGqc.
33 Ron Katims, "Vietnam," undated and unpublished manuscript in possession of the author.
34 Katims, "Vietnam."
35 Katims, "Vietnam."
36 National Museum of American History, "Sea-Land Intermodal Transport," http://amhistory.si.edu/onthemove/collection/object_969.html.

Chapter 5 From the Ocean to the Docks

1 For more on the history of New York Harbor and the Hudson River, see Arthur G. Adams, *The Hudson through the Years* (Westwood, NJ: Lind, 1983).
2 Marian Betancourt, *Heroes of New York Harbor: Tales from the City's Port* (Guilford, CT: Globe Pequot, 2017), 90–107.
3 "Ambrose Channel Tested," *Zion's Herald*, September 18, 1907.
4 "Sandy Hook Light to Be Dropped To-Day," *New York Times*, November 30, 1908.
5 "Lightship Ambrose," *Ship Spotting World*, http://ship.spottingworld.com/Lightship_Ambrose.
6 Werner Bamberger, "Coast Guard Seeks to Replace 2 Lightships," *New York Times*, March 8, 1964.
7 "Tower to Replace Harbor Sentinel," *New York Times*, August 22, 1967.
8 "Tanker Hits Ambrose," *Hull Truth Boating Forum*, https://www.thehulltruth.com/boating-forum/152381-tanker-hits-ambrose-light.html.
9 Evan True, "New York's Ambrose Tower Hit by Ship—Again," *Professional Mariner*, March 2, 2007.
10 Dom Yanchunas, "Tanker Damages Ambrose Light, Raising Questions about Its Future," *Professional Mariner*, January 16, 2008.
11 Dom Yanchunas, "Ambrose Light, Deemed Obsolete, Passes into History," *Professional Mariner*, November 24, 2008.
12 Public Affairs Department, U.S. Coast Guard, "Ambrose Light to Be Deconstructed after 41 Years of Service" (press release, July 25, 2008).
13 For more information on the historic forts of New York and New Jersey, see Robert B. Roberts, *Encyclopedia of Historic Forts: The Military, Pioneer, and Trading Posts of the United States* (New York: Macmillan, 1988).
14 Greg Hitchen, interview, USCG Sector New York, Fort Wadsworth, Staten Island, NY, July 12, 2018.
15 Sandy Hook Pilots Association, "Pilotage Service," https://www.sandyhookpilots.com/PilotageService.aspx.
16 Aileen Weintraub, *Navesink Twin Lights* (New York: Rosen, 2003), 5–15.

17. Thomas Hoffman, *Fort Hancock* (Charleston, SC: Arcadia, 2007), 7–8; and John R. Weaver II, *A Legacy in Brick and Stone: American Coastal Defense Forts of the Third System, 1816–1867* (Missoula, MT: Pictorial Histories, 2001), 117–119.
18. New Jersey Lighthouse Society, "Sandy Hook Light," http://www.njlhs.org/njlight/sandy.html.
19. Randall Gabrielan, *Sandy Hook* (Charleston, SC: Arcadia, 1999), 10.
20. New Jersey Lighthouse Society, "Sandy Hook Light."
21. David Veasey, *Guarding New Jersey's Shore: Lighthouses and Life-Saving Stations* (Charleston, SC: Arcadia, 2000), 21.
22. Preservation New Jersey, "Romer Shoal Light," https://www.preservationnj.org/listings/romer-shoal-light/.
23. Veasey, *Guarding New Jersey's Shore*, 22–23.
24. Lighthouse Friends, "West Bank Lighthouse," https://www.lighthousefriends.com/light.asp?ID-757.
25. Lighthouse Friends, "West Bank Lighthouse."
26. Linda Dianto, Executive Director of the National Lighthouse Museum, telephone conversation with the author, September 13, 2019.
27. Veasey, *Guarding New Jersey's Shore*, 23–24.
28. Lighthouse Friends, "Old Orchard Shoal Lighthouse," https://www.lighthousefriends.com/light.asp?ID=758.
29. Dianto telephone conversation.
30. Reuven Blau, "Coney Island's Parachute Jump Gets $2 Million Upgrade and 8,000 LED Lights," *New York Daily News*, April 11, 2013.
31. Lighthouse Friends, "Coney Island Lighthouse," https://lighthousefriends.com/light.asp?ID=394.
32. Lighthouse Friends, "Coney Island Lighthouse."
33. M. O. Poole, "Historic Islands at New York's Front Door," *New York Times*, February 28, 1937.
34. Gerald R. Wolfe, "Hoffman Island," in *The Encyclopedia of New York City*, ed. Kenneth T. Jackson (New Haven, Conn.: Yale University Press, 1995), 549.
35. Gerald R. Wolfe, "Swinburne Island," in Jackson, *Encyclopedia of New York City*, 1146.
36. Poole, "Historic Islands."
37. John P. Murphy, "Training the Seaman," *New York Times*, June 29, 1941.
38. Joyce Lam, "The Abandoned Man-Made Islands in NYC: Hoffman and Swinburne Islands," *Untapped Cities*, July 29, 2013, https://untappedcities.com/2013/07/29/hoffman-and-swinburne-islands-abandoned-man-made-islands-in-new-york-city.
39. Rose Jimenez, "Gravesend Bay: A Geographic Case Study" (Earth and Environmental Sciences, Graduate Center, City University of New York), https://www.arcgis.com/apps/MapTour/index.html?appid=c93d234dd0824791aa4fa7314a23c7d8.
40. Sandy Hook Pilots, *New York-New Jersey Harbor, Hudson River, and Long Island Sound* (Staten Island, NY: Sandy Hook Pilots, 2014), 23.
41. Greg Hitchen, Vessel Traffic Service Director at Coast Guard Sector New York, telephone conversation with the author, September 13, 2019.
42. A complete and informative account of the Verrazzano-Narrows Bridge can be found in Gay Talese's remarkable book *The Bridge,* originally published in 1964, when Talese was a reporter for the *New York Times*. A paperback edition with a new preface and a new afterword was published in 2003. Talese renders a sympathetic account of the lives of the ironworkers who did the work. He describes the

construction of the towers, the spinning of the cables, the building of the roadway, the installation of the lighting, and the painting of the ironwork. See Gay Talese, *The Bridge* (Brooklyn, NY: Walker, 2003).

43 Brian Merlis and Lee Rosenzweig, *Brooklyn's Bay Ridge and Fort Hamilton* (Brooklyn, NY: Israelowitz, 2000), 25.
44 Sandy Hook Pilots, *New York–New Jersey Harbor*, 8.
45 Frederick R. Black, *A History of Fort Wadsworth, New York Harbor* (Boston: National Park Service, 1983), 5.
46 North American Forts, "Coast and Harbor Defenses of New York City," https://www.northamericanforts.com/East/nycity3.html#harbor.
47 National Park Service, "Detailed History of Fort Wadsworth," https://www.nps.gov/gate/learn/historyculture/ftwad.htm.
48 Merlis and Rosenzweig, *Brooklyn's Bay Ridge*, 13; and Weaver, *Legacy in Brick and Stone*, 111–113.
49 Military Bases, U.S. Army, "Fort Hamilton," http://www.militarybases.us/army/fort-hamilton/.
50 "Anchorage Regulations: Port of New York," *Federal Register*, November 16, 2006, https://www.federalregister.gov/documents/2006/11/16/E6-19314/anchorage-regulations-port-of-new-york.
51 Gregory P. Hitchen, Vessel Traffic Service Director at Coast Guard Sector New York, correspondence with the author, September 26, 2019. Hitchen cited the implementation of the 48-hour time limit from the *Federal Register* 41, no. 167 (August 26, 1976).
52 NOAA Office of Coast Survey, "NOAA Adds Grid Overlay to Chart Anchorage Areas of Port of New York and New Jersey" (April 26, 2018), https://noaacoastsurvey.wordpress.com/2018/04/26/noaa-adds-grid-overlay-to-chart-anchorage-areas-in-port-of-new-york-and-new-jersey/.
53 For more information on the ferry, see the nicely illustrated Staten Island Museum, *Staten Island Ferry* (Charleston, SC: Arcadia, 2014).
54 New York City, Department of Transportation, "Staten Island Ferry About," https://www.siferry.com/ferry-about.html.
55 Gerard R. Wolfe, "Kill van Kull," in Jackson, *Encyclopedia of New York City*, 636.
56 Pamela Silvestri, "Atlantic Salt Company Put Its Mark on the Staten Island Waterfront over the Years," *Staten Island Advance*, May 16, 2018, https://expo.silive.com/erry-2018/05/c1690048c96615/index.html.
57 International-Matex Tank Terminals, "Bayonne Terminal," https://www.imtt.com/terminal-locations/bayonne/.
58 Simon Zorovich, Senior Harbor Pilot, telephone interview with the author, October 1, 2019.
59 Barnett Shepherd, "Sailor's Snug Harbor," in Jackson, *Encyclopedia of New York City*, 1032.
60 Caddell Dry Dock and Repair Company, "Over 100 Years of Service," http://www.caddelldrydock.com.
61 Wendell Jamieson, "Tugboat Alley," *New York Times*, August 21, 2005.
62 The deckhouses are painted in the official colors of "purple red," RAL 3004, and the hulls are painted in "chrome green," RAL 6020. RAL is a standardized European matching system that defines colors for paint. See RC Groups, "Correct Colors for Moran Tugs," https://www.rcgroups.com/forums/showthread.php?2550536-Correct-colors-for-Moran-tugs.

63. Moran Towing Corporation, "Corporate History," https://www.morantug.com/site/#/corporatehistory.
64. Moran Towing Company, "Fleet Lists," https://www.morantug.com/site/#/home.
65. Nautical Class, "Maritime Examinations Guide," http://nauticalclass.com/what-are-the-types-of-tugs-and-use.
66. Nautical Class, "Maritime Examinations Guide."
67. Budworth Marine, "Articulated Tug Barge," https://www.vesselrepair.com/articulated-tug-barge.html; and Daniel Zorovich, telephone interview with the author, October 5, 2019.
68. Tim Heffernan, "A Bridge Too Low," *Atlantic*, April 2015.
69. WSP Construction Company, "Bayonne Bridge Raising Opens Ports to World's Largest Ships," https://www.wsp.com/en-US/insights/bayonne-bridge-raising-opens-ports-to-worlds-largest-ships.
70. WSP Construction Company, "Bayonne Bridge Raising."
71. Kate Archer, "Going Up! A Bridge Makes Way for Bigger Ships," *New York Times*, March 21, 2014.
72. Hugh R. Morley, "Largest Ship to Call US East Coast Passes under Bayonne Bridge," *Journal of Commerce*, July 17, 2017.

Chapter 6 Navigation

1. "The Most Dangerous Job in the Military," YouTube, https://www.youtube.com/watch?v=Ut5UonKZ_gk.
2. U.S. Coast Guard Forum, "Buoy Tender," http://www.uscg.org/Forum/aft/9451.aspx.
3. George E. Kreitemeyer, *The Coast Guardsman's Manual* (Annapolis, MD: Naval Institute Press, 2000), 42–43.
4. Jeremy D'Entremont, *The Lighthouse Handbook: Hudson River & New York Harbor* (Kennebunkport, ME: Cider Mill Press, 2009), 96–104.
5. Lieutenant Torrey Jacobsen, USCG, interview with the author, Bayonne, NJ, February 25, 2020.
6. Jacobsen interview.
7. Jeff Cowan, "Eliminating Aids to Navigation?," *Maritime Logistics Professional Magazine* 2Q (2014): 54, https://magazines.marinelink.com/Magazines/MaritimeProfessional/201405/content/eliminating-aids-navigation-471981.
8. Cowan, "Eliminating Aids to Navigation?"
9. Jacobsen interview.
10. Jacobsen interview.
11. Jacobsen interview.
12. Jacobsen interview.
13. Jacobsen interview.
14. Jason Wiley, interview with the author, Bayonne, NJ, September 18, 2012.
15. Wiley interview.
16. Wiley interview.
17. Jason Willey, interview with the author, New York Harbor, October 16, 2012; and Jim Hendricks, "Navigating at Night," *Boating Magazine* (n.d.), http://www.boatingmag.com/skills/seamanship/navigating-night?page=0,2.
18. Greg Hitchen, interview with the author, Staten Island, NY, July 12, 2018.
19. Hitchen interview.

20 Hitchen interview.
21 Hitchen interview.
22 Hitchen interview.
23 Hitchen interview.
24 Hitchen interview.
25 Carrieann Dixon, interview with the author, Staten Island, NY, July 12, 2018.

Chapter 7 Pilotage

1 In referring to the pilots, I have used abbreviated terminology. The formal name for the New York group is "United New York Sandy Hook Pilots' Benevolent Association." The formal name for the New Jersey group is "United New Jersey Sandy Hook Pilots' Benevolent Association." For more on the history of these groups, see Captain Edward C. Winters Jr., *Northwest 3/4 West: The Experiences of a Sandy Hook Pilot, 1908–1945*, ed. Mel Hardin (Staten Island, NY: United New Jersey Sandy Hook Pilots' Benevolent Association & United New York Sandy Hook Pilots' Benevolent Association, 2004).
2 William Wertenbaker, "The Sandy Hook Pilots," *New Yorker*, September 14, 1968.
3 Again, I have used abbreviated terminology in the interest of brevity in the text. The formal name for the New York group is "The Board of Commissioners of Pilots of the State of New York." The formal name for the New Jersey group is "The New Jersey Maritime Pilot and Docking Pilot Commission."
4 Sandy Hook Pilots Association, "Fleet," https://www.sandyhookpilots.com/Fleet.aspx.
5 Nick Cutmore is the Secretary General of the International Maritime Pilots Association (IMPA). See "Safe Hands: The Role of the Maritime Pilot," *Ship Technology*, https://www.ship-technology.com/features/featuresafe-hands-the-role-of-the-maritime-pilot-4574416/.
6 Captain John Oldmixon informed me by telephone on January 10, 2020, that his great-grandfather was Ivan Van Albert, born in Sweden in 1840. His grandfather was Lionel "Lee" Oldmixon, and his father was John Oldmixon I.
7 Captain John Oldmixon informed me by telephone on January 10, 2020, about some of his close relatives who have also served as pilots. These include R. L. Oldmixon, his brother; Andrew Oldmixon, his father's cousin; and Kenneth Sorenson, his mother's twin brother.
8 For more on the continuing place of tradition in American culture, see Simon J. Bronner, *Following Tradition: Folklore in the Discourse of American Culture* (Logan: Utah State University Press, 1998), 9–21.
9 Captain John Oldmixon, interview with the author, Bridgewater, NJ, October 3, 2019.
10 Oldmixon interview.
11 Oldmixon interview.
12 "Safe Hands."
13 Oldmixon interview.
14 Oldmixon interview.
15 John Oldmixon, interview with the author, Bridgewater, NJ, January 25, 2020.
16 John Oldmixon, telephone interview with the author, January 10, 2020.
17 Oldmixon telephone interview.
18 Oldmixon telephone interview.

19   James L. Kolstad, National Transportation Safety Board, "Safety Recommendation," Log M-371-B, April 18, 1991.
20   Simon Zorovich, telephone interview with the author, April 14, 2020.
21   Ronal Smothers, "Treacherous Shoals . . . of Trenton? Harbor Pilots Divided Over Plan to Regulate Them," *New York Times*, June 16, 2004.
22   Simon Zorovich, interview with the author, Matawan, NJ, July 25, 2019.
23   For more on the increase in size of container ships in this period, see Brian J. Cudahy, *Box Boats: How Container Ships Changed the World* (New York: Fordham University Press, 2006), and Marc Levinson, *The Box: How the Shipping Container Made the World Smaller and the World Economy Bigger* (Princeton, NJ: Princeton University Press, 2006).
24   Jean-Paul Rodrigue, *The Geography of Transport Systems*, 5th ed. (New York: Routledge, 2000).
25   Marine Insight News Network, "HMM Names World's Largest Container Vessel, 24,000 TEU Giant, 'HMM Algeciras' at DSME Shipyard," *Marine Insight*, https://www.marineinsight.com/shipping-news/hmm-names-worlds-largest-container-vessel-24000-teu-giant-hmm-algeciras-at-dsme-shipyard/.
26   Zorovich interview.
27   Zorovich interview.
28   Simon Zorovich, telephone interview with the author, April 29, 2020.
29   This docking arrangement is known as four and two. That means that we have four headlines and two spring lines forward, plus four stern lines and two spring lines aft.
30   Zorovich interview.

Chapter 8   Tugboats

1   The sons of senators are roughly 8,500 times more likely to become senators than the average American man. See Joe Pinsker, "Like Father, Like Son: How Much Nepotism Is Too Much?," *Atlantic*, June 11, 2015.
2   Quoctrung Bui and Claire Cain Miller, "The Jobs You're Most Likely to Inherit from Your Mother and Father," *New York Times*, November 22, 2017.
3   Simon Zorovich, interview with the author, Matawan, NJ, August 20, 2016.
4   Paul Fieldhouse, "Eating Together: The Culture of the Family Meal," *Vanier*, September 18, 2019.
5   Daniel Zorovich, interview with the author, Matawan, NJ, August 25, 2016.
6   For more on the long history of McAllister Towing in the New York–New Jersey harbor, see Patricia Keefe, "McAllister Towing," *Maritime Reporter & Engineering News*, November 2016.
7   Daniel Zorovich interview.
8   Daniel Zorovich, telephone interview with the author, April 18, 2020.
9   Fred Ferretti, "To Tugman This Cook Is the Best," *New York Times*, September 14, 1983.
10   Daniel Zorovich interview, August 25, 2016.
11   The U.S. Coast Guard established an inspection system for towing vessels at Chapter 46 of the Code of Federal Regulations, Subchapter M. It was published in the *Federal Register* on June 20, 2016. See Towing Vessel Inspection Bureau, https://www.thetvib.org.
12   For more on the history and development of the Z-drive, see George Matteson, *Tugboats of New York* (New York: New York University Press, 2005), 230–233.

13. Daniel Zorovich interview, August 20, 2016.
14. [Simon/Daniel] Zorovich interview.
15. [Simon/Daniel] Zorovich interview.
16. [Simon/Daniel] Zorovich interview.
17. It is clear from the literature that, for state pilots, the federal pilot license issued by the U.S. Coast Guard is the national minimum standard. All of the coastal states, including New Jersey, have additional requirements. For more on this issue, see Paul G. Kirchner, "A Career as a Ship Pilot," *Coast Guard Proceedings*, Fall 2008.

Chapter 9 The Contemporary Port

1. U.S. Army Corps of Engineers, "Fact Sheet—New York and New Jersey Harbor Deepening," February 1, 2019, https://www.nan.usace.army.mil/Media/Fact-Sheets/Fact-Sheet-Article-View/Article/487407/fact-sheet-new-york-and-new-jersey-harbor-50-ft-deepening/.
2. "NY-NJ Port Completes 50-Foot Channel Project," *Journal of Commerce*, September 2, 2016, https://www.joc.com/port-news/us-ports/port-new-york-and-new-jersey/ny-nj-port-completes-50-foot-channel-project_20160902.html.
3. Panama Canal Museum, "Canal Expansion," https://cms.uflib.ufl.edu/pcm/timeline/expansion.aspx.
4. Steven Mufson, "An Expanded Panama Canal Opens for Giant Ships," *Washington Post*, June 26, 2016, https://www.washingtonpost.com/world/the_americas/an-expanded-panama-canal-opens-for-giant-ships/2016/06/26/11a93574-37d1-11e6-af02-1df55f0c77ff_story.html.
5. WSP, "Bayonne Bridge Raising Opens NJ Ports to World's Largest Ships," https://www.wsp.com/en-US/insights/bayonne-bridge-raising-opens-ports-to-worlds-largest-ships.
6. WSP, "Bayonne Bridge Raising."
7. Hugh R. Morley, "NY-NJ Handles Mega-Ships Post-Bayonne Bridge in Stride," *Journal of Commerce*, June 18, 2018, https://www.joc.com/port-news/us-ports/port-new-york-and-new-jersey/ny-nj-handles-mega-ships-post-bayonne-bridge-lift-stride_20180618.html.
8. Hugh R. Morley, "NY-NJ Lays Ground for 18,000-TEU Ships," *Journal of Commerce*, 28 December 2018, https://www.joc.com/port-news/port-productivity/ny-nj-lays-ground-18000-teu-ships_20181228.html-0.
9. Bethann Rooney, interview with the author, Port Newark, NJ, October 8, 2019.
10. Bill Mongelluzzo, "Keeping Current: Ports Use Information Portals to Propel Data Sharing," *Journal of Commerce*, January 6, 2020.
11. Bill Mongelluzzo, "Productivity Shortfalls Compromise Ports' Ability to Service Mega-Ships," *Journal of Commerce*, December 16, 2019.
12. Rooney interview.
13. Mike Frassinelli, "Exit 13-A: A Trip along the 'Most Dangerous Two Miles in America,'" *Star Ledger*, March 31, 2019.
14. David E. Rosenbaum, "Agency Chiefs Readjusting to the New Imperatives, *New York Times*, September 11, 2002.
15. DHS, "Creation of the Department of Homeland Security," https://www.dhs.gov/creation-department-homeland-security.
16. State of New Jersey, "Office of Homeland Security and Preparedness," https://www.njhomelandsecurity.gov/.

17. Anderson Cooper Blog, "The Most Dangerous Two Miles in America," August 15, 2006, http://www.cnn.com/CNN/Programs/andersoncooper.360/blog/2006/08/most-dangerous-two-miles-in-america.html.
18. Rooney interview.
19. Maria Galluci, "At Last, the Shipping Industry Begins Cleaning Up Its Dirty Fuels," *Yale Environment 360*, June 28, 2018, https://e360.yale.edu/features/at-last-the-shipping-industry-begins-cleaning-up-its-dirty-fuels.
20. McKinsey & Company, "Bunker Fuel," *Energy Insights*, n.d., https://www.mckinseyenergyinsights.com/resources/refinery-reference-desk/bunker-fuel/.
21. Fred Pearce, "How 16 Ships Create as Much Pollution as All the Cars in the World," *Daily Mail*, November 21, 2009, https://www.dailymail.co.uk/sciencetech/article-1229857/How-16-ships-create-pollution-cars-world.html.
22. Galluci, "At Last."
23. Queensland Government, "Sulfur Dioxide," n.d., https://www.qld.gov.au/environment/pollution/monitoring/air/air-pollution/pollutants/sulfur-dioxide.
24. Pearce, "How 16 Ships Create."
25. Galluci, "At Last."
26. Bill Mongelluzzo, "Recouped Interim Low-Sulfur Surcharges Bode Well for Carriers," *Journal of Commerce*, December 4, 2019.
27. Rooney interview.
28. In my interview with Bethann Rooney on October 8, 2019, she explained that the Port Authority of New York and New Jersey has a remarkably good record in meeting environment policy challenges not just in the maritime sphere, but also in the operation of tunnels and bridges as well as at terminals, the rail system, the aviation department, and the World Trade Center. The Port Authority is committed to a 35 percent reduction by 2025 and an 80 percent reduction by 2050. She also pointed out that the Port Authority has signed on to the Paris Climate Accord.
29. Rooney interview.
30. Rooney interview.
31. Bill Rooney, "Executive Commentary," *Journal of Commerce*, January 6, 2020.
32. Rooney interview.
33. Rooney interview.
34. Rooney interview.
35. National Science and Technology Council, *Intermodal Cargo Transportation: Industry Best Security Practices* (Cambridge, MA: U.S. Department of Transportation, 1999), 9.
36. "Cargo Theft: A Billion-Dollar Problem," *Ship Technology*, July 30, 2019.
37. Lee Willet, "Cargo Theft Report Suggests Risk Lies Ashore Rather Than at Sea," *Safety at Sea*, July 10, 2019.
38. Rooney interview.
39. National Science and Technology Council, Intermodal Cargo Transportation, 18.
40. Alex Wright, "Everything You Should Know about Cargo Theft and How to Prevent It," *Risk and Insurance*, February 18, 2019.
41. Wright, "Everything You Should Know."
42. Rachel Abrams, "Taking a Peek into Santa's Bag," *New York Times*, December 20, 2014.
43. Abrams, "Taking a Peek into Santa's Bag."
44. Customs and Border Protection, "CBP New York/Newark Seizes Nearly $1.7 Million in Counterfeit Nike Sneakers," https://www.cbp.gov/newsroom/local

-media-release/cbp-new-yorknewark-seizes-nearly-17-million-counterfeit-nike-sneakers.
45 United Nations Office on Drugs and Crime, "Container Control," https://www.unode.org/unodc/en/urban-safety/container-control.html.
46 National Science and Technology Council, Intermodal Cargo Transportation, 13.
47 John Sullivan and Henry DeGeneste, *Policing Transportation Facilities* (Springfield, IL: Charles C Thomas, 1994), 90.
48 Sullivan and DeGeneste, *Policing Transportation Facilities,* 92–94.
49 Port Authority of New York and New Jersey, "Containerized Cargo," https://www.panynj.gov/port/containerized-cargo.html.
50 Rooney interview.
51 Rooney interview.
52 The widespread damage of Hurricane Sandy has been well documented. Consider, for example, a book of 127 pages published by the *New York Post*, titled simply *Sandy*, with nearly 100 full-color images giving an overview of the damage in Manhattan, Brooklyn, Queens, Staten Island, New Jersey, Long Island, and the region (Chicago: Triumph Books, 2013). A more comprehensive account is the book of 359 pages written by Kathryn Miles with the title *Superstorm: Nine Days Inside Hurricane Sandy* (New York: Penguin, 2014). The author writes a wide-ranging account of the hurricane with many fascinating stories, including an account of the sinking of the replica ship HMS *Bounty* off the coast of North Carolina on October 29, 2012. However, for our purposes, we shall focus on the storm's impact on Port Newark and Port Elizabeth.
53 University Transportation Research Center, "Final Report: Lessons from Hurricane Sandy for Port Resilience" (New York: City College of New York, 2013), abstract, 5.
54 University Transportation Research Center, "Final Report."
55 Rooney interview.
56 Rooney interview.
57 Kate Sheppard, "Report Warns That Superstorm Sandy Was Not 'the Big One,'" *Huffington Post*, December 6, 2017.
58 Winnie Hu, "Americans Are on a Shopping Spree. These Workers Are Overwhelmed," *New York Times*, March 31, 2021.
59 Hu, "Americans Are on a Shopping Spree."

Chapter 10   Moving the Freight

1 I first learned of this four-way simplifying concept by watching the YouTube video by Michael Hurlock, "Rubber Tired Gantry Training, Volume 1, RTG," https://www.youtube.com/watch?v=YK-kNyOmVpg.
2 For more on these concerns, see Chuqian Zhang, Jiyin Liu, Yat-wah Wan, Katta G. Murty, and Richard J. Linn, "Storage Space Allocation in Container Terminals," *Transportation Research Part B: Methodological* 37 (2003): 883–903, https://www.sciencedirect.com/science/article/abs/pii/S0191261502000899.
3 See Robert Stahlbock and Stefan Voss, "Operations Research at Container Terminals: A Literature Update," *OR Spectrum* 30 (2008): 1–52, https://link.springer.com/article/10.1007/s00291-007-0100-9.
4 MGAR, "Sea-Land History," http:www.mgar.net/sealand2.htm.

5. Giovanni Antonuccio went on to explain in an interview on January 28, 2020, at Port Elizabeth that the Maersk Group is the holding company that owns not only the Maersk Line and APM Terminals but also the steamship lines Sea Land, Safmarine, and Spitzer.
6. Giovanni Antonuccio, interview with the author, Elizabeth, NJ, January 28, 2020.
7. "N4 It's Time for More," https://www.navis.com/globalassets/brochures/brochure_navis-n4.pdf.
8. Antonuccio interview.
9. Giovanni Antonuccio, email correspondence with the author, July 31, 2020.
10. Antonuccio interview.
11. RBS, "BAPLIE," https://www.rbs-emea.com/glossary-entry/baplie.html.
12. As this was written, APM Terminals had six Post Panamax Cranes, four Super Panamax Cranes, and Four Ultra-Super Panamax Cranes according to a "Terminal Information" sheet provided to the author by Giovanni Antonuccio on August 5, 2020.
13. To better understand how shoreside gantry cranes work, I found it useful to view Michael Hurlock, "Container Crane Training Vol. 1," https://www.youtube.com/watch?v=jU25QoNiowU.
14. Giovanni Antonuccio, telephone conversation with the author, August 5, 2020.
15. In this book, I have stuck with the customary industry vocabulary of "longshoreman" and "longshoremen" as people employed in a port to load and unload ships. Clearly the current terms are for the male gender. Of course, women are capable of doing the job, and I have met several of them. None of them objected to the terms; rather, they embraced them. At the risk of being politically incorrect, I have used the terms that are commonly used in the industry. In this context, I should point out that the labor union on the West Coast has gotten around the problem of male gender preference by adopting the name "International Longshore and Warehouse Union" (ILWU).
16. Giovanni Antonuccio, telephone conversation with the author, August 7, 2020.
17. Antonuccio telephone conversation, August 7, 2020.
18. Antonuccio telephone conversation, August 7, 2020.
19. Antonuccio telephone conversation, August 7, 2020.
20. Nico Berx, "Efficient Stacking on Container Terminals, How to Cope with Volume Fluctuations," *Port Technology*, February 18, 2011, https://www.porttechnology.org/technical-papers/efficient_stacking_on_container_terminals_how_to_cope_with_volume_fluctuati/.
21. Razouk Chafik, Youssef Benadada, and Jaouad Boukachour, "Stacking Policy for Solving the Container Stacking Problem at a Containers Terminal" (University of Bordeaux, Sixth International Conference of Information Systems, Logistics, and Supply Chain, 2016).
22. Hugh R. Morley, "New ILA System Aims to Boost NY-NJ Port Performance," *Journal of Commerce*, March 4, 2019.
23. For more on the life of these truck drivers, see "7 Advantages to Becoming a Long Haul Truck Driver," *Digital Boom*, January 15, 2019, https://adigitalboom.com/7-advantage-to-becoming-a-long-haul-truck-driver.
24. Edward Mack, "10 Best Country Truck Driving Songs of All Time," *Wide Open Country*, n.d., https://www.wideopencountry.com/10-best-truckin-songs-country-music/.

25 Alex Pappas, "What Is Drayage?," *Supply Chain Dictionary*, January 25, 2012, https://dedola.com/2012/01/what-is-drayage/.
26 For a readable history of trucking in New Jersey, see Clifford B. Ross, *The Motor Carriers of New Jersey* (East Brunswick: New Jersey Motor Truck Association, 1969).
27 Dick Jones, telephone interview with the author, August 1, 2018.
28 Association of Bi-State Motor Carriers, https://www.bistatemotorcarriers.com.
29 Jones telephone interview.
30 Jones telephone interview.
31 Bill Mongelluzzo, "Categorical Imperative: Trucking Industry Seeks Clarity on Driver Classification Issues," *Journal of Commerce*, January 6, 2020.
32 Mark Szakonyi, "Classify This," *Journal of Commerce*, January 20, 2020.
33 Szakonyi, "Classify This."
34 Best Transportation, "Owner-Operators and S863. What Does It Mean for Trucking in New Jersey?," n.d., https://besttrans.com/owner-operators-and-s863-what-does-it-mean-for-trucking-in-new-jersey/.
35 Joni Casey, "Executive Commentary," *Journal of Commerce*, January 6, 2020.
36 Philip Gigante, telephone interview with the author, September 26, 2019.
37 Carolyn Said, "AB5 Cleanup Bill Allows Gig Work for Musicians, Interpreters, More in California," *San Francisco Chronicle*, September 4, 2020, https://www.sfchronicle.com/business/article/AB5-clean-up-bill-allows-gig-work-for-musicians-15528850.php.
38 Kerry Jackson, "Assembly Bill 5 Is Already Destroying Jobs and Opportunities," *Orange County Register*, December 31, 2019, https://www.ocregister.com/2019/12/31/assembly-bill-5-is-already-destroying-jobs-and-opportunities/.
39 Said, "AB5 Cleanup Bill."

Chapter 11   The Seamen's Church Institute

1 Douglas Stevenson, interview with the author, Port Newark, NJ, September 6, 2012.
2 Leah Robinson Rousmaniere, *Anchored within the Veil: A Pictorial History of the Seamen's Church Institute* (New York: Seamen's Church Institute of New York and New Jersey, 1995), 14–16.
3 Rousmaniere, *Anchored within the Veil*, 18–21.
4 Rousmaniere, *Anchored within the Veil*, 23.
5 Rousmaniere, *Anchored within the Veil*, 35–37, 49–51.
6 Rousmaniere, *Anchored within the Veil*, 47.
7 Thomas W. Ennis, "Seamen's Haven Put Up for Sale," *New York Times*, June 5, 1965.
8 Rousmaniere, *Anchored within the Veil*, 92.
9 Lawrence O'Kane, "Seamen Will Get a 23-Story Home," *New York Times*, October 20, 1966.
10 Anthony Depalma, "Replacing Seamen's Institute," *New York Times*, September 18, 1965.
11 Rousmaniere, *Anchored within the Veil*, 122.
12 Richard F. Shepard, "Ahoy, Mates, the Institute's Back at the Seaport," *New York Times*, May 6, 1991.
13 Shepard, "Ahoy, Mates."
14 Jontham Sederstrom, "Seamen's Institute to Sell Its Building and Leave Manhattan," *New York Times*, October 5, 2010.

15 Sederstrom, "Seamen's Institute."
16 No container ships, tankers, break-bulk ships, nor car carriers dock in Manhattan at this time; however, the New York Passenger Ship Terminal is active. It is a terminal for oceangoing passenger ships located at piers 88, 90, and 92, along Twelfth Avenue on the Hudson River between West 46th and West 54th.
17 Rev. David M. Rider, telephone conversation with the author, May 6, 2020.
18 Stevenson interview.
19 SCI fundraising letter sent to the author, April 29, 2014.
20 Michael Kimmel, *Manhood in America: A Cultural History* (New York: Free Press, 1996), 295.
21 Rev. Megan Sanders, interview with the author, Port Elizabeth, NJ, October 2, 2012.
22 Second Officer Darko Gardic, interview with the author, Port Newark, NJ, October 2, 2012.
23 Second Officer Yahya Musbaiton, interview with the author, Port Newark, NJ, October 2, 2012.
24 Sanders interview, October 2, 2012.
25 Rev. Megan Sanders, interview with the author, Port Elizabeth, NJ, October 9, 2012.
26 Second Officer Jean-Marie Franceschi, interview with the author, Port Elizabeth, NJ, October 9, 2012.
27 First Officer Blot Guilhem, interview with the author, Port Elizabeth, NJ, October 9, 2012.
28 Ibid.
29 Sanders interview, 9 October 2012.
30 Rev. James Kollin, interview with the author, Port Newark, NJ, June 8, 2021.
31 Rev. Cora DiDomenico, interview with the author, Port Newark, NJ, June 8, 2021.

Chapter 12 The Future

1 Port Authority of New York and New Jersey, advertisement, *Journal of Commerce*, September 16, 2019.
2 My optimistic outlook on the future of Port Newark–Elizabeth is based, in part, on the "Port Master Plan 2050" released by the Port Authority of New York and New Jersey in 2019. It is a detailed, ambitious, and comprehensive document that lays out a plan for the entire port in both states including Bayonne, Staten Island, Brooklyn, and Manhattan.
3 I borrowed this old saying from Janet Nodar, who quoted it in her article "Lessons from the Front Row" in an editorial from the *Journal of Commerce* special issue on "Breakbulk & Project Cargo," published in June 2020.
4 William Safire and Leonard Safir, eds., *Good Advice* (Avenel, NJ: Wings Books, 1992).
5 Costas Paris, "Coronavirus Slows but Won't Halt Shipping's Focus on Global Trade," *Wall Street Journal*, May 1, 2020.
6 Michael Angeli, "NY-NJ Port Seeks Aid amid COVID-19 Volume Decline," *Journal of Commerce*, May 22, 2020.
7 Alan M. Field, "The US-China Trade War: Where We Go from Here," a white paper released by the *Journal of Commerce*, part of HIS Markit, a business information company, November 2019.

8. Field, "US-China Trade War."
9. Mark Szakonyi, "Cyberattack on Maersk Highlights Need for Global Strategy," *Journal of Commerce*, June 28, 2017.
10. Alana Semuels, "Fewer Jobs, More Machines," *Time*, August 17–24, 2020.
11. Hugh R. Morley, "NY-NJ Coalition to Enhance Cyber Security," *Journal of Commerce*, January 3, 2020.
12. Semuels, "Fewer Jobs, More Machines,"
13. Bill Mongelluzzo, "More North American Port Automation Expected," *Journal of Commerce*, July 4, 2019.
14. Peter Tirschwell, "Sooner Rather Than Later," *Journal of Commerce*, March 16, 2020.
15. Tirschwell, "Sooner Rather Than Later."
16. Tirschwell, "Sooner Rather Than Later."
17. Dan Kopf, "Union Membership in the US Keeps on Falling, Like Almost Everywhere Else," *QuartzThings*, February 5, 2019, https://qz.com/1542019/union-membership-in-the-us-keeps-on-falling-like-almost-everywhere-else/.
18. Nandita Bose and Krystal Hu, "How Big Unions Smooth the Way for Amazon Worker Protests," *Technology News*, May 21, 2020, https://www.reuters.com/article/us-health-coronavirus-amazon-com-workers/how-big-unions-smooth-the-way-for-amazon-worker-protests-idUSKBN22X19Q.
19. Alex Kingsbury, "The Two Companies Have Been Pleading Their Case to the House and Senate," *U.S. News and World Report*, December 8, 2009.
20. Christopher Mims, "As E-Commerce Booms, Robots Pick up Human Slack," *Wall Street Journal*, August 8, 2020.
21. "Port Automation Technology Demonstrated by Cosco," *Maritime Executive*, May 13, 2020, https://maritime-executive.com/article/port-automation-technology-demonstrated-by-cosco.
22. Peter Tirschwell, "ILWU Move on LA Automation 'a Slippery Slope,'" *Journal of Commerce*, July 10, 2019.
23. Bill Mongelluzzo, "More North American Port Automation Expected," *Journal of Commerce*, July 4, 2019.
24. "Moody's: Port Automation May Not Always Deliver Results," *Maritime Executive*, August 12, 2020, https://www.maritime-executive.com/article/moody-s-port-automation-may-not-always-deliver-results.
25. For more on the relationship between ports and ships, see Ira Breskin, *The Business of Shipping* (Atglen, PA: Schiffer, 2018).
26. "Loss Prevention—Maritime Accidents Top Reasons," *Maritime Cyprus*, August 11, 2020, https://www.maritimecyprus.com/2020/08/11/loss-prevention-maritime-accidents-top-reasons/.
27. Mike Schuler, "Rena Grounding Final Report Reveals Errors, Lack of Oversight," *gCaptain*, December 18, 2014, https://gcaptain.com/rena-grounding-final-report/.
28. "Stricken Cargo Ship Rena Breaks Up Off New Zealand," *BBC News*, January 8, 2012, https:www.bbc.com/news/world-asia-16458574.
29. Kristy Johnston, Karla Akuhata, Angela Cuming, and Michael Daly, "Sinking Rena Leaking Oil," *Waikato Times*, January 10, 2012, https://archive.is/20120914012410/http://www.stuff.co.nz/waikato-times/news/6232985/Split-Rena-sinking.
30. Schuler, "Rena Grounding Final Report."

31. "Jury Still Out on Hanjin Pennsylvania Fire," *Hazcheck*, May 14, 2003, https://hazcheck.existec.com/imo-publications/general.asp?np=news_61.
32. Ruddy Branka, Guy Marlair, and Brigitte Nedelec, "Risk Analysis of Fireworks in Cargo Container Ships" (International Symposium on Fireworks, Porto, Portugal, October 2010), https://www.researchgate.net/publication/278771100_Risk_analysis_of_fireworks_transport_in_cargo_container_ships.
33. "Jury Still Out on the Hanjin Pennsylvania Fire."
34. "Hanjin Pennsylvania," *Cedre*, September 14, 2009, https://wwz.cedre.fr/en/Resources/Spills/Spills/Hanjin-Pennsylvania.
35. Chase Samuels, "MOL Comfort Accident: The Worst Shipping Disaster in History," *Trader Risk Guaranty*, July 17, 2019, https://traderiskguaranty.com/trgpeak/mol-comfort-worst-shipping-disaster/.
36. Rob Almeida, "ClassNK Completes Forensic Study of MOL Comfort Structural Failure," *gCaptain*, September 30, 2014, https://gcaptain.com/mol-comfort-investigation-report-released/. Please note that this article provides a link to the full 123-page technical report.
37. Luke Smout, "Hull Fracture Caused MOL Comfort Sinking," *Felixstowe Docker*, http://www.felixstowedocker.com/2014/10/hull-fracture-caused-mol-comfort-sinking.html.
38. "Autonomous Minesweeper Set for 'Live' Operations," *Maritime Executive*, January 14, 2020, https://www.maritime-executive.com/article/autonomous-minesweeper-set-for-live-operations.
39. Robert Quigley, email correspondence with the author, August 13, 2020.
40. Dana Merkel, "Opportunities and Challenges with Maritime Autonomous Surface Ships (MASS)," *Maritime Reporter*, October 10, 2019.
41. "Terra Sound Completes First NOAA U.S. Charting Survey Using an Autonomous Surface Vehicle" (press release, ASV Global, August 16, 2015), https://www.asvglobal.com/terrasond-completes-first-noaa-u-s-charting-survey-using-an-autonomous-surface-vehicle/
42. Stave Dimitropoulos, "Will Ships without Sailors Be the Future of Trade?," *BBC Business News*, July 16, 2019, https://www.bbc.com/news/business—48871452.
43. Dimitropoulos, "Will Ships without Sailors Be the Future of Trade?"
44. Robert Quigley, email correspondence with the author, February 20, 2020.
45. "Sea Trials Begin for Mayflower Autonomous Ship's 'AI Captain,'" (press release, IBM Edge Computing, March 5, 2020), https://newsroom.ibm.com/2020-03-05-Sea-Trials-Begin-for-Mayflower-Autonomous-Ships-AI-Captain.
46. David Dubay, "Why We Will Never See Fully Autonomous Commercial Ships," *Maritime Executive*, June 25, 2019, https://www.maritime-executive.com/editorials/why-we-will-never-see-fully-autonomous-commercial-ships.
47. "New Network Supports Port Readiness for Autonomous Shipping," *Maritime Executive*, August 6, 2020, https://www.maritime-executive.com/article/port-network-to-address-challenges-and-support-autonomous-shipping.

Index

Page numbers in italics indicate photographs.

AED (automated external defibrillator), 147
AI captain, 225
Aids to Navigation Team (ANT), 103–10
air pollution, 158, 159–60; and conversion of cranes to electric power, 169; reduced by lower speed, 160
air traffic, near NJ Turnpike Exit 13-A, 157
AIS (automatic identification system), 111, 112–13, 116
Alaska Freight Lines, 62–64
Allen, Nancy, 24
allision with Ambrose Light, 72–73
Amazon, 161, 215
Ambrose, John Wolfe, 70–71
Ambrose (LV-111), 71
Ambrose (WLV-613), 71
Ambrose Channel, 71, 143; stationing of Pilot Boat 1 or 2 at beginning of, 116
Ambrose Light, 71
Ambrose Light Vessel (LV-87), 71, *72*
Ambrose Offshore Light Station, 71–73
Ambrose Sea Buoy, 74
America (launch), 116
America Export Lines, 40
American Bureau of Shipping, 42
Anchorage Port Authority, 63

Anderson, William, 27
antiaircraft batteries: Fort Hamilton, 83; Fort Hancock, 75
Antonuccio, Giovanni, 174–75, 247n5
A.P. Moller-Maersk Group, 174, 201
APM Terminals, 167, 172, 173–80, 247n5; cranes, *174,* 247n12; Port Elizabeth, 6
ARCIMS (Atlas Remote Combined Influence Minesweeping System), 222
Armed Services Vocational Aptitude Battery (ASVAB), 104
Army Corps of Engineers, 71; and dredging of main channels in Port of New York and New Jersey, 151–52, 154; and expansion of Cam Ranh Bay harbor, 64
army depot, Port Newark, 12
Army Materiel Command, 65
articulated tug-barge combinations (ATBs), 88, 89
artificial intelligence, 214, 215
ASCs (automated stacking cranes), 217
Association of Bi-State Motor Carriers, 153, 182, 185, 188
Astrolabe Reef, 218, 219
ASV (autonomous surface vessels); ASV *Global, 222,* 223–24
Atlantic, Gulf, and Pacific Company, 55

253

Atlantic Salt Company, 86
ATON (Aids to Navigation), 4; maintenance, 92–103. *See also* buoys; daymarks
automated external defibrillator (AED), 147
automatic identification system (AIS), 111, 112–13, 116
automation, 6; of container ships, 221–26; at Hamburg container terminal, *216*; increased, and ILA-NYSA agreement, 180; of minesweeping, 222; of transportation industry, Moody's Investor Service on, 214, 217
automobile processors, 155
Ava McAllister, tug, 141, *142*
azimuth thruster, 89. *See also* Z-drive

Bader, Jeff, 153
Baguio, Capt. Francisco, *208*
balance, shipboard, of container weights, 182–83
Ballantine, Peter, 9
bank, seamen's, at SCI South Street hotel, 193
BAPLIE (bayplan including empties), 177
barge transportation, 181
Barlow, suction dredge, 55
bar pilots, 1, 5. *See also* Sandy Hook pilots
Bartholomew, Harland, 26–27
Battle of the Atlantic, 23
Battle of the Wilderness, 83
Bauman, Milton P., 21
Bayonne Bridge, 17, 89, *89, 91;* Navigational Clearance Project, 152–53; raising of roadway, 5
bayplan including empties (BAPLIE), 177
BBT Logistics, 187
Beauregard, Pierre Gustave Toutant, 48
Belgian block, xv
Belzer, Michael, 186
Bennett, David, 159
Bentley, Helen Delich, 183
Bergen Point, 91
Besson, Gen. Frank S., Jr., 65–66
Best Transportation, 186, 187
Bienville, Jean-Baptiste Le Moyne de, 47
BMW, 155
Board of Commissioners of Pilots of the State of New York, 242n3
boat utility stern loaders (BUSLs), 106

"bomb carts," 177
Bone, Capt. Craig, 80, 122
Bonner, Herbert C., 46
bottlenecks, port, 161–62
bow thrusters, lacking in Navy ships, 148
Brady, John A., 27
break-bulk cargo handling, 32; discharge at Cam Ranh Bay harbor, 65; *vs.* containerizing, 49, 50; and theft, 162–63
Brennan, Prof. Lawrence, 224
Bridge, The (Talese), 239–40n42
Broad Street and Merchants Association, Newark, 28
Brown Industries, 41
Bruce A. McAllister, twin screw tugboat, *91*, 139–40, *140*
BT Nautilus, runs aground, 121
Bulk import productivity comparison, 155
"Bunker C" fuel, 158
buoys, 92; can, 108; for Kill van Kull dredging project, 106; nun, 108; sea (*see* sea buoys); servicing of, 101–3
buoy tenders, 93–103, *94, 102*; "keeper class," 95; steering, *94*
Busan (Pusan), South Korea, 203
Bush, Irving T., 15
Bush, Pres. George H. W., 79
Bush, Pres. George W., 156
Byrne, Joseph, 27

Caddell Dry Dock and Repair Company, 87–88
calcium hypochlorite, 220
call sizes, 155
Cam Ranh Bay, Vietnam, harbor expansion at, 64–68
Canas, Richard, 157
cargo handling equipment, and pollution, 160
cargo ships: C-3 and C-4 types, 44; C-2 type, 31–32, 43, 44, 53; World War I, 13–14
cargo terminal, Port Elizabeth, 59
Carlin, Leo P., 35, 37
Casey, Joni, 187
Centralized Examination Station (CES), 185
Central New Jersey Railroad, acquisition of tidal marsh site from, 36

CES (Centralized Examination Station), 185
chain, 108
chain of command, on tugboats, 146
chaplains, SCI, 190–91, 198; and pastoral care, 209; seamen's rapport with, 201, 206; ship visits by, 206
chrome ore, 30
Chubb Insurance, 163
Churchill, Winston, 212
Clancy, William J., 28–29
ClassNK, 221
CMA CGM *Argentina*, 210
CMA CGM *Tosca*, 201, 202–4
coastal artillery batteries, Fort Wadsworth, 83
Coast Guard: administers Transport Workers Identification Credentials, 5 (*see also* TWIC); closes Port of New York and New Jersey during Hurricane Sandy, 168; in cyber security coalition, 213; deaccessions West Bank Lighthouse, 78–79; fog regulation, 130; imposes 48-hour limit for anchoring, 84, 240n51; inspection for towing vessels, 243n1; licensing by, 116, 139, 149–50, 244n17; maintains Aids to Navigation (ATON), 4, 92–103; National Vessel Movement Center, 73; Office of Navigations Systems, 98; Oldmixon collision hearing, 119–20; Regional Examination Center, Battery Park, 148; search and rescue missions, 92–93, 97; Sector New York, 73; tests *Ideal X*, 42; tugboat regulations, 143; Vessel Traffic Service (VTS), 4, 85; and "war on drugs," 166
Coast Guard Academy, 96–97, 110
Coast Guard Auxiliary, 112
cocaine, 165, 166
Colden, Cadwallader David, 15–16
commercial vessels, and pollution, 160
Compagnie Générale Transatlantique, 204
computers, SCI, seamen's access to, 190, *190*, 197
Coney Island, 79
Coney Island Lighthouse, 79
congestion, terminal, *161*, 161–62, 184; at Port Elizabeth, 170; at Port Newark, 170; at Port of Long Beach, *161*
Conrad, Joseph, 194

Consumer Products Safety Commission, 164–65
containerization, 2, 37; during Vietnam War, 67, 68–69; expansion of, 3–4; and inspections on port exit, 157; Malcom McLean's role in, *38,* 38–50; stacking (*see* stacking); standardization of sizes, 50–51
Container Terminal Altenwerder (CTA), Hamburg, *216*
container weight, 182–83
container yard, 173, 178–79
contraband in containers, 164–66, 185, 186
Cook Inlet, Alaska, 63
Corbett, James, 159
Cornell, Scott, 164
Cosco Shipping Ports Limited, 216
Cosco Shipping Rose, container ship, 216
Costello Dismantling Company, 73
Council on Port Performance (CPP), 162
counterfeit merchandise, 165
COVID-19 pandemic, 170–71, 188, 208, 209, 212; USNS *Comfort*, in response efforts, 122
cranes, 173; automated stacking (ASCs), 217; converted to electric power, 169; height of, raised to accommodate larger ships, 154; loading SS *Ideal X, 44*; Maersk's post-Panamax upgrading to, 175, 247n12; onboard SS *Gateway City,* 44–45; at Port Newark Container Terminal, *167*; shipboard, 53–54; ship-to-shore (STS) container, 177, 179; shore, 54; weight limits, 182; yard, 179
crew rotation, 206–7; and COVID-19 pandemic, 210
criminals, ASV vulnerability to, 225
crimping, 192
Crossing under the Hudson: The Story of the Holland and Lincoln Tunnels (Gillespie), x
CSS *Alabama*, 47–48
C-2 cargo ships, 53
Cullman, Howard S., 30
Cunningham, John T., 9, 19–20
Cushing, Charles R., 3, 43–44, 53
Customs and Border Protection (CPB), 165, *166*, 185; and "war on drugs," 166
Cutmore, Nick, 116, 119, 242n5

Cutrale Citrus Products, xv
Cyber Advisory Committee, 213
cyberattack, 213–14
cyberattacks, 6
Cyberkeel, 213

Dacey, Tim, 120–21
Daggett, Harold J., 180
dangerous cargo, 218
Day, Capt. Michael H., 85
daymarks, 92
deadheading, 50
DeAngelo, Beth, 90, 153
deckhands, tug, 144
DeLong, Col. Leon B., 64
DeLong piers, 64–68
Dempsey, Raymond J., 21
Department of Homeland Security, 156
destroyer escorts (DEs), built at Port Newark, 23
destroyers, built at Port Newark, 22
Dewey, Commodore George, 74
Dianto, Linda, 78–79
DiDomenico, Rev. Cora, 207–10, *210*
Dixon, Carrieann, 113–14
DK Shipping Company, Rotterdam, 158
dock fenders, 133
docking pilots, 1, 5, 87, 115; licensing of, 120–22, 149–50; procedure, 130; role of, 122–30; tugboat captains as, 137; tugboat communications with, 144; vessel handoff from Sandy Hook pilots, 144
docking pilots' association, membership after apprenticeship, 128
dockside cranes, 177
Doig, Jameson W., 15, 16, 26
Doubleday, Gen. Abner, 83
draft, channel, 128
drayage, 181, 186; contractor vs. employee model, 186–88; trucks, as pollution source, 160
dredging, 5; of Ambrose Channel, 71; Depression-era federal funding, 21; of Kill van Kull, 106, 154; of main navigation channels in Port of New York and New Jersey, 151–52; of meadowlands channel, 10; Passaic River, paid for by federal funding, 19; Port Elizabeth, 55, *152*;
postwar deepening of Port Newark channel, 30; of San Juan port, 61
Drexel, Richard J., 47
Driscoll, Alfred E., 28
Drug Enforcement Agency (DEA), 165; and "war on drugs," 166
drug smuggling, 165–66; interdiction of, 97, 110
dry dock services, 87
drying closet, seamen's, at SCI South Street hotel, 193
Dubay, Commander David USCG, 225–26
dunnage, 32, 33; reclamation, 34
dynastic jobholding, 117, 135, 137–38, 242nn6–7; among senators, 243n1

earthquake, Alaskan, of 1964, 63–64
"East Bound and Down" (song), 181
ECDIS (Electronic Chart Display and Information System), 98, 116
EDT (estimated department time) of container, 180
efficiency, and terminal size, 167–68
Eifler, Gen. Charles William, sold on containerization by Ron Katims, 66–67
ELD (electronic logging device), 184
electric winches, and mooring lines, 133
Electronic Chart Display and Information System (ECDIS), 98, 116
electronic logging device (ELD), 184
Elizabeth Marine Terminal, 1, 51, 54, 174; cranes at APM facility, *174*. *See also* Port Elizabeth
Ellen McAllister, tractor tug, 140, *141*
Emergency Fleet Corporation, 13
employee *vs.* contractor model, in drayage trucking, 186–88
engineers, tugboat, 144–45
engine order telegraph (EOT), 123
engine speeds, 123
environmental concerns: over main-channels dredging, 151–52; over port activities, 158, 159–60; and Port Authority compliance, 245n28
EPIRB (emergency position indicating radio beacon), 147
escort towing, 88–89

estimated department time (EDT) of container, 180
explosions, 220
eyeglasses, seaman's, replacement of, 209
EZ-Pass, 214

F-A-C, 109
FAPS (automobile processors), 155
Fast Aids-to-Navigation Response Team (FART), 109
FBI, and "war on drugs," 166
Federal Shipbuilding and Dry Dock Corporation, 22
Fed-Ex, 215–16
ferries, and pollution, 160
FEUs (forty-foot equivalent units), 178
Fieldhouse, Prof. Paul, 137–38
finger piers, 87
firefighting, 222
fires, shipboard, 218, 220
fireworks, maritime transport of, 220
fisheries enforcement, 110
fixed-price bidding, Sea-Land's, for Vietnam work, 67
Floating Church of Our Saviour for Seamen, 192
floods, 6; Hurricane Sandy, 168
Florio, Gov. James, 120–21
fog, 129–30
food: chaplains share onboard meals with crews, 208; lunch aboard the *Tosca,* 202; onboard, tugboat, 142–43
Fort Hamilton, 83–84
Fort Hancock, 75
Fort Richmond, 83
Fort Schuyler, 139
Fort Tompkins, 83
Fort Wadsworth, *82,* 83; Coast Guard Sector Command Center and Vessel Traffic Service at, 73
"four and two" mooring line arrangement, 243n29
Franceschi, Second Officer Jean-Marie, 202
Frank Grad and Sons, 57
Franklin Lumber Company, 29
Fresnel lenses, 74, 76
Fruehauf Trailer Company, 48
Fuller, Burton, 21

Gardic, Darko, 199–200
gate complex, 173; upgraded at APM Terminal, 175
Gateway National Recreation Area, 75, 81
General Services Administration (GSA), auctions off West Bank and Old Orchard Shoal Lighthouses, 78
Gigante, Philip, 187–88
gig economy, 188
Gillette Company, 48
Gilroy, John, 142
Glassing, Capt. Andrew E., 90–91
Global Container Terminal, 172, 217
Global Express Service, 159
Goethals Bridge, 17, 126
Gold, Jonathan, 213
Governors Island, 103
GPS (Global Positioning Satellites), 92, 95, 98–99
Gravesend Bay Anchorage, 81
Great Depression, effect on Newark, 19–20
Greco, Frank, 32, 34
greenhouse gases, reduction in emissions of, 160
grounding of vessels, 121, 129, 218–19, *219*
Guilhem, First Officer Blot, 202–3, 204
gypsum, xv

hackers, ASV vulnerability to, 225
hailer, 144
Hancock, Gen. Winfield Scott, 75
Hanjin Pennsylvania, container ship, 220
harbor pilots, 82–83
Heart of Darkness (Conrad), 194
heaving line, 133
Heimgartner, Tom, 186–187
herbor pilots. *See* docking pilots
High Bridge Stone Company, xv
Hitchen, Gregory P., 110–13, 240n51
HMM (Hyundai Merchant Marine), 127
HMM *Algeciras,* 127
HMS *Bounty* replica ship, sinking of, 246n52
Hoffman, John T., 80
Hoffman Island, 80–81
"hogging," 221
holdmen, 178
Hong Kong, 203
Horizon Shipbuilding Incorporated, *142*

horizontal transport, 173
Hours of Service (HOS), 184
House Merchant Marine Committee, 46
Hublein Spirits, 48
Hudson, Henry, 70
Hugh, Rob, 225
Hughes, Gov. Richard J., 55, 174
human error, 218
Hurley, Edward N., 14
Hurricane Sandy, 77, 78, 109, 168–70, 246n52
Hushcraft, 224
"hustlers," 177–78
hydrauling drums, and mooring lines, 133
Hyundai Merchant Marine (HMM), 127

ICC lighterage case, 15–17, 20–22
ILWU (International Longshore and Warehouse Union), 217
imports, surge of during COVID-19 pandemic, 171
Industrial Exhibition of 1872, 9
infantry riflemen, U.S. Army, 93
infectious disease, 212. *See also* COVID-19 pandemic
integrated tug-barges (ITBs), 89
Intermodal Association of North America, 187
intermodal carriers, 181–82
International Association of Marine Aids to Navigation and Lighthouse Authorities, 226
International Association of Ports and Harbors, 226
International King's Point Cargo Handling Exposition (1961), 51
International Longshore and Warehouse Union (ILWU), 217, 247n15
International Longshoremen's Association (ILA), 46, 50, 168, 178, 180, 212, 217
International Maritime Organization (IMO), 158–59, 222–23, 226
International-Matex Tank Terminals (IMTT), Bayonne, 86–87
International Seafarers' Center, Port Newark, 6, *190*, 194, *196*, 197, 202, *205*
internet connection for seamen, on SCI computers, 202
Isla Grande Airport, Puerto Rico, 60

Jackson, Gen. Thomas "Stonewall," 83
Jacobs, Bradley, 214–15
Jacobsen, Lieut. Torrey, 95–98; commands *Katherine Walker,* 100–103; on hurricanes, 99–100
Jamieson, Wendell, 88
Jarka, Capt. F., 19
Jensen, Lars, 213
Jersey Central Railroad, 29
Jersey Gardens Outlet Mall, 191, 198, 202, 206, 207
Johns, R. Kenneth, 41, 49
Johnson, Seaman Chance, *96*
Jonathan C. Moran, Z-drive tug, 140–41
Jones, Dick, 182
Jones Act, 3, 42; and Alaska trade, 62; Port Elizabeth cargo initially covered by, 58, 59, 60

Katims, Ron, 3, 4, 62; and Alaska earthquake relief, 63; and the designing of Port Elizabeth, 56–58; and new port development, 60, 61; sells Gen. Charles William Eifler on containerization, 66–67
Keats, Alan, 68
"keeper class" buoy tenders, 95
Kelly, Ed, 122
Kill van Kull, 86, *89,* 111, 127–28; additional dredging proposed, 154; buoys, 106
King, Lyle, 53
Kollin, Rev. James, 206–7, *208*
Kolp, Hattie, 171
Korwatch, Capt. Lynn, 99
Kouras V freighter, 72
Kramer, Charles F., 18

LaFarge Gypsum, xv
Land, Emory Scott, 31
LaRoe, Wilbur, Jr., 16
launches, for pilot transportation, 116
law, and maritime automation, 222–23, 224
LeBlanche, Nick, 195
Lee, Gen. Robert E., 83
Lehigh Valley Railroad, 28, 29
Le Moyne, Jean-Baptiste, Sieur de Bienville, 47
Leone Pancaldo, Italian destroyer, 136
Liberty ships, 31, 32
lifesaving equipment, tugboat, 147

Light Emitting Diodes (LED) on sea buoys, 73
lighterage, 14; ICC rulings, 15–17, 20–22
lighthouses: construction, 77–78. *See also individual lighthouses*
lighting, Port Elizabeth, 57
Lincoln, Pres. Abraham, ix
Linkin, Dr. Megan, 170
locomotives, and pollution, 160
logbooks: truckers', 184; tugboat, 146
longshoremen, 128, 178, 247n15
longshore unions, 7, 215, 216–17
Looking for America on the New Jersey Turnpike (Gillespie), x
Lord Jim (Conrad), 194
Lorier, Jim, 195
Lowe, Donald V., 37
LSFO (low-sulfur fuel oil), 159, 160
luggage storage, seamen's, at SCI South Street hotel, 193

MacArthur, Gen. Douglas, 31
Maersk Line, 174–75, 247n5; 2017 cyberattack, 213
Maersk SeaLand, 174
Maher Terminal, 167, 172
mail service, seamen's, at SCI South Street hotel, 193
Manhattan, catamaran, 192
Mansfield, Archibald Romaine, 192–93
Marconi, Guglielmo, 74
marine carpenters, 32–34
marine depot, Port Newark, 34–35
marine operations building, Port Elizabeth, 58
maritime academies, 138. *See also individual academies*
Maritime Association of New York, 121
maritime autonomous surface ships (MASS), 226
Maritime Commission, Federal. *See* U.S. Maritime Commission
maritime school, Hoffman Island, 80–81
Maritime Studies program, Williams College, 189
Markowitz, Marty, 79
marshalling yards: Isla Grande Airport, Puerto Rico, 60; Port Elizabeth, 57

MASS (maritime autonomous surface ships), 226. *See also* ASV
Massachusetts Maritime Academy, 119
Mayaguez, Puerto Rico, 61
Mayflower, ASV catamaran, 225
McAllister Towing & Transportation, 123, 126, *133,* 137
McAllister yard, *91, 133*
McCormack, William J., 17
McGreevey, Gov. Jim, 122
McLean, Malcom, 2, 3, 37, *38,* 38–50, 235n4; buys T-3 tankers, 51; meets with Gen. Frank S. Besson, Jr., 65–66; preference for chassis operation, 57; submits fixed-price bid for Vietnam work, 67, 68–69
McLean Trucking Company, 39–40, 45
McLeod, Capt. Charles L., 55
Mediterranean Shipping Company (MSC), 199
mental health, seafarers', 210
Merchant Marine Act of 1920, 42. *See also* Jones Act
Mercur Trading Company, 19
Meyner, Gov. Robert B., 35, 36, 54–55
migrant interdiction, 97, 110
Military Ocean Terminal, Bayonne, 104
Military Sealift Transportation Service, 67
Miller, Troy, 165
mine sweeping, automation of, 222
"Mister Big" (William J. McCormack), 17
"Mister Potato" animated cartoon (Port Authority), 17
Mitsui OSK Lines, 220
MOL *Comfort,* 220–21
Møller, Arnold Peter, 174, 201
Moller, Christian Eyde, 158
money: exchange, SCI, 195; transfer service, SCI, 6, 190, *190,* 191, 208
Mongelluzzo, Bill, 155
Moody's Investor Service, on container terminal automation, 214, 217
mooring crew, 132–33
mooring lines, 132
Moran, Michael, 88
Moran Towing, 140; tugboat yards, 88–89, 126
Morrow, Edward A., 51
Moscoso, Theodoro, 60–61
Moses, Robert, 82

Motor Carrier Act of 1935, 40
motor carriers, 181–88
motorsailers, 136
MSC *Silvana,* 199
M/T *Aegeo,* 72
M/T *Axel Spirit,* 72
Murphy, Vincent J., 27–28
Musbaiton, Second Officer Yahya, 200
M/V *Maersk Singapore, 123, 125*
M/V *Oleander,* 195
MV *Rena,* grounding of, *219*
Mystery Maiden figurehead, 195

Nabisco, 48
Nardi, John, 180, 212
Narrows, the, 82. *See also* Verrazzano Narrows Bridge
National Aids to Navigation School, 97
National City Bank of New York (later Citibank), 3, 45
National Lighthouse Museum, 76–77
National Retail Federation, 213
National Vessel Movement Center, 73
Navesink Twin Lights, 74
Navis N4, 175
Newark, city of: creation of deepwater port, 9 (*see also* Port Newark); finances deepening of Newark Bay channel, 18; founding and early history of, 8–9; manufacturing at, 9; 1929 stock market crash, 19–20
Newark, passenger steamboat, 9
Newark airport, postwar capital needs, 25
Newark Association of Commerce and Industry, 56
Newark Bay Channel, depth of, 19
New Jersey (Pilot Boat 2), 116
New Jersey Citizens Highway Committee, 56
New Jersey Maritime Pilot and Docking Pilot Commission, 149–50, 242n3; Oldmixon collision hearing, 119–20
New Jersey Office of Homeland Security and Preparedness, 156
New Jersey Turnpike: generates pollutants, 160; vulnerability around Exit 13-A, 156, *156*
Newman, Tom, 224
New York (Pilot Boat 1), 116

New York City Department of Transportation, 86
New York Naval Station, 83
New York Passenger Ship Terminal, 249n16
New York Shipping Association (NYSA), 180, 212; and Council on Port Performance, 162
New York State–New Jersey ICC suit, 14–16, 15–16
Nike missile antiaircraft systems: Fort Wadsworth, 83; Sandy Hook, 75
Ningbo, China, 203
nitrogen oxide, 159
Nodar, Janet, 249n3
Norfolk and Western Railroad, 41
Norton, Lilly and Company, Port Newark terminal, 35–36
NotPerya cybervirus, 213
NSA (National Security Agency), 197

Office of Coast Survey, National Oceanic and Atmospheric Administration (NOAA), 85
Office of Naval Research (ONR), 24
Office of Navigations Systems, Coast Guard, 98
oil; "Bunker C" fuel, 158; exports to Finland, 30; storage tanks and refineries, near NJ Turnpike Exit 13-A, 157, *157*
Oldmixon, Andrew, 242n7
Oldmixon, Capt. John II, 116–18, 242n6; and reduced apprenticeship period, 120–21
Oldmixon, John I, 242n6
Oldmixon, Lionel "Lee," 242n6
Oldmixon, R.L., 242n7
Old Orchard Shoal Lighthouse, 78–79
on-dock railroad yard, 173
"on duty while driving," 184–85
online shopping, and surge in container traffic, 170
On the Waterfront (film), 36
OOCL *Berlin,* 90–91
Operational Risk Management Procedure, 109
"Operation Bootstrap," Puerto Rico, 60
operations research, in terminal management, 173
O'Reilly, John Boyle, 48

Orient Overseas Container Line (OOCL), 90
Outerbridge, Eugenius Harvey, 15
Outerbridge Crossing, 17
owner-operator truckers, 186

Pacific Maritime Association (PMA), 217
Panama Canal, 62, 90; expansion of, 5, 152–53
Pan-Atlantic Steamship Corporation, 42, 48, 49
pandemics, 6; fatigue/burnout from, 170
Parachute Jump, Coney Island, 79
Paré, Michael, 24
Paris Climate Accord, 245n28
Parkinson, Dr. Brad, 98–99
particulate matter, air pollution from, 159
pastoral care, 206
Pauly, Chief Warrant Officer Darren, 73
pay: of docking pilots, 125–26; pooled salaries, of pilot association members, 115–16, 150; of Sandy Hook pilots, 121
Pearce, Fred, 158
penetration testing, 213
Peninsula at Bayonne Harbor, the, 104
Pennsylvania Railroad, 28, 29
Phantom (launch), 116
Philadelphia Experiment, 23–24
Philadelphia Experiment, The (film), 24
Philippines, as Sea-Land's staging area for Vietnam supplies, 68
Phillips 66, 126
phone cards, low-cost, SCI, 190, 199
pier, 173
pilferage, 163, 166. *See also* theft, minimization of
pilotage, 4–5
pilot boats, 116, 118–19
pipelines, near NJ Turnpike Exit 13-A, 157
piracy: in Southeast Asian waters, 163; risk of, in Indian Ocean, 203; Vietnam-era, 68
pleasure boats, 112
PODs (ports of destination), 180
Pomare, 225
Ponce, Puerto Rico, 61
Port Authority of New York: assumes management of Port Newark, 3; creation of, 16–17; headquartered in NJ, ix; leases northeast channel land and warehouses from U.S. Navy, 29; negotiates Port Elizabeth lease and costs with Sea-Land, 55, 56
Port Authority of New York and New Jersey: Bayonne Bridge roadway raising project, 90; and Council on Port Performance, 162; and leasing of Port Newark Container Terminal, *167*; offers pollution-reduction incentives, 160; productivity concerns, 155; *vs.* environmental groups, about main-channels dredging, 152
port charges, 62
Port Elizabeth, 55, *59*; APM Terminal (*see* APM Terminals); buildings, 57–59; cargo terminal, 59; congestion at, *161*; designing of, 56–59; development announced, 36; dredging, *152*; leases facilities to Sea-Land, 55; marine operations building, 58
port governance, 154–55
Port Houston, offloading area at, 53
Portland-California Steamship Company, 20
"Port Master Plan 2050" (Port Authority of New York and New Jersey), 249n2
Port Newark: channel depth, 19; congestion at, *161*; creation of, 9–11; during Korean War, 30–34; map, *ii*, 2; marine depot, 34–35; Port Authority takeover of, 25–29; post-Korean War growth spurt, 34–36; proximity to NJ Turnpike, 156, *156*; in the Roaring Twenties, 17–19; SCI International Seafarers' Center, 6, *190*, 194, *196*, 197, 202, *205*; World War I expansion of, 11–14; in World War II, 22–24
Port Newark Container Terminal, 167, *167*, 172
Port of Baltimore, 183
Port of Long Beach, 62; congestion at, *161*
Port of New York and New Jersey: closed by Coast Guard during Hurricane Sandy, 168; container terminals, 5–6; cyber security coalition, 213
Port of Oakland, 62
Port of Rotterdam Authority, 226
Port of San Francisco, 62
Port Reading, 168
ports: of destination (PODs), 180; disruptions to, 212–13; future opportunities, 214–18; new, Sea-Land constructs, 61

Potero Hills. See SS *Ideal X*
power outages, 6
Princess Bay, 78
productivity: increases in, ILA-NYSA agreement on, 180; of Port Authority ports, 155
Proper Protective Equipment (PPE), 201
Puerto Rican Ports Authority (PRPA), 61
Puerto Rico: Jones Act cargo, 59, 60; truck transportation assumes primacy, 61
punctuality, tugboat crew, 145

quay crane, 179

radar, 113
railroads: across meadowlands, 10; antagonism with truckers, 41; connections Port Newark's former Navy warehouses, 29; freight rates and lighterage, 15, 20–21; on-dock yard, 173; oppose Port Authority takeover of Port Newark, 28; Port Authority tunnel proposals, 16, 17; well suited for intercity transport, 181
RAL, 240n62
Randall, Robert Richard, 87
ransomware attacks, 213
recreational boats, 120
Red Hook Container Termial, 172
Reed, Jerry, 181
Reilly, Sheridan, 78
repair, of autonomous vessels, 224–25
rescue swimmers, 93
"restricted waters," 102
RFID (radio frequency identification) truck tags, 175, 182
Richardson, Paul F., 40, 51–52
Rider, Rev. David, 195, *196*
RMS *Caronia*, 71
Robert E. McAllister, tug, 148–49
Robins Reef Light, 95
Romer Shoal Lighthouse, 76–77
Rooney, Bethann, 154, 157, 159, 160, 167, 171, 245n28; on lessons of Hurricane Sandy, 169–70; on port congestion, 161–62
RTG (rubber-tired gantry) cranes, 179
Rutgers University, cyber security experts from, 213

Safmarine steamship line, 247n5
"sagging," 221
Sailor's Snug Harbor, 87
Sailor's Town, Manhattan, 191
Sanderson, Megan, 197–201, 204
Sandy Hook Lighthouse, *75, 76*
Sandy Hook Light Vessel (LV-51), 71
Sandy Hook Peninsula, 74–75
Sandy Hook pilots, 74, 82, 95, 114, 115–22, 143; apprenticeship period, 120–21; association of, 117; New York *vs.* New Jersey, 115–20; numbers, 117–18; vessel handoff to docking pilots, 144; working board, 118
Sandy Hook Pilots Association, 121
Sasebo Dock Yard, 65
SBC. *See* Submarine Boat Corporation
Schubert, Frank, 79–80
SCI. *See* Seamen's Church Institute
SC *Marigot*, 207, *208*
scrap metal exports to Philippines, 30
Seaboard Airline, railroad, 41
sea buoys, 72; Ambrose, 74
Seafarers' Club, 195
seafloor mapping, 223–24
Sea-Land Service, 2, *44,* 49, 51; becomes dominant carrier for Puerto Rico, 60; constructs new container ports, 61; leases Port Elizabeth facilities, 55; sold to Maersk, 174
Sea Land steamship line, 247n5
Seamen's Church Institute (SCI), xiii–xiv, 6, 197–210; International Seafarers' Center, *190,* 194; origins, 191–92; South Street Seaport building, 194–95, *196*; State Street building, Manhattan, 194
Seamen's Home, 192
search and rescue missions, 92–93, 97
Seattle, 203
Secret Service, and "war on drugs," 166
Sector New York, Coast Guard, 73
security, post 9/11, 155–57
self-insurance, shippers, against theft losses, 163
Semmes, Raphael, 47–48
Seward, Alaska, port of, 63
Shanghai, 203
shared meals, and children's socialization, 137–38

Sharp, George C., 44
Shedd, John A., v
Sherrerd, Morris R., 11
ship arrival scheduling, 176
shipboard cranes, 53–54
shipbuilding, at Port Newark: refitting C-2 cargo ships during Korean War, 31–32; World War I, 12–14; World War II, 22–24
ship disruptions, 218–21
ship opportunities, future, 221–26
ship-to-shore (STS) cranes, 177
shipwrecks, 218
shopping: proxy, by chaplains, 191, 195, *210*; shop-at-sea, 207, *208*
shore cranes, 54; shoreside gantry cranes, 177
"shot" (chain length), 108
showers, seamen's at SCI South Street hotel, 193
Siegel, Michael (cartographer), *ii*
SIM cards, cell phone, 6, 208
Simpson, Ben, 224
Singapore, recruitment of labor from, for Sea-Land's Vietnam work, 68
sinkers, 108
"situational awareness," 99
Skype, at International Seafarers' Center, 197
small craft, 112
Smith, J. Spencer, 15
Smith, Jean R., xiii–xiv, 190
Smith, Seaman Jessica, *100*
Smit International, 221
smuggling of drugs, 165–66
Sorenson, Kenneth, 242n7
speed: and increased risk, 129; reduction of, and lessened pollution, 160
spill response, 222
Spitzer steamship line, 247n5
spring lines, 132
SS *Agawam*, 13–14
SS *Almena*, 42
SS *Azalea City*, 47
SS *Beauregard*, 48
SS *Bienville*, 47
SSCGC *Tahoma*, 110
SS *Elizabethport*, 51
SS *Esso New Orleans*. See SS *Elizabethport*
SS *Examelia*, 40

SS *Fairland*, 48
SS *Gateway City*, 44–45, 46
SS *Ideal X*, 2, 37, *38*, *39*, 42, *43*, 44, *44*, 48
SS *Normandie*, 204, *205*
SS *Raphael Semmes*, 47, 49, 50
stacking, 57, 173, 179–80, 183
Stakem, Thomas E., 50–51
standardization: of container sizes, 50–51; of fittings and hardware, 51; Stapleton Anchorage, 84–85
Staten Island Ferries, 85–86; St. George and Whitehall Terminals, *85*
Statue of Liberty, 86
steel, Swedish, 30
steering, of Navy ship by tug at base of bow, 148
Stevens Institute of Technology, cyber security experts from, 213
storms, 6; hurricanes and ATON, 99–100; Superstorm Sandy, 77, 78, 109, 168–70, 246n52; Tropical Storm Noel, 72
structural failures, of ships, 218, 220–21
structure discrepancies, 105–6
Submarine Boat Corporation (SBC), 12–13, 14, 20, 22
Suez Canal, 203
sulfur dioxide emissions, 158, 160
sulfuric acid, 158
SUNY Maritime College, 139
Supreme Court, U.S.. See U.S. Supreme Court
Sutpen, Henry R., 14
Svendsen, Sigurd, 193
Sweatshops on Wheels: Winners and Losers in Trucking Deregulation (Belzer), 186
Swift and Company, 20
Swinburne, John, 80
Swinburne Island, 80

Tama, Capt. Jason, 214
Tammany Hall, 17
Tanjung Pelepas, Malaysia, 203
tankers: T-2, 42; T-3, 51
Tantlinger, Keith, 3, 41, 42
Tarden, Capt. James, 110
Tarnef, Barry, 163
Teamsters, 186
telephone service, international seamen's, 190

Talese, Gay, 239–40n42
tenders, buoy, 93–103
terminals: anatomy of, 173; size of, 167–68. *See also individual terminals*
TerraSond, 223–24
terrorists, ASV vulnerability to, 225
TEUs (twenty-foot equivalent units), 127, 178
theft, minimization of, 162–64
tides, and docking windows, 126, 128
Titanic, SCI memorial lighthouse beacon, 193
TNT Express, 213
Tobin, Austin J., 26, 27–28, 37
Torrijos, Pres. Martín, 152
Tosca (Puccini), 202
"to-to" trailer shipping, 41
Toyota, 155
tractor tugboats, 89
trade wars, 6, 212–13
traffic fluctuations, 126, 127
trailerable aids to navigation boats (TANBs), 106–7
Transmarine Lines, 20
transportation, by SCI chaplains, 199, 206, 208
Transportation Security Administration (TSA), xiv
Transport Workers Identification Credential (TWIC), xiv, 198, 201
Trent, Robert, 8
Trinity Church, Wall Street, and establishment of SCI, 191
T. Roosevelt, Panamax ship, 153
Tropical Storm Noel, 72
trucks: appointment system, 176; over-the-road drivers, 181
truck stops, as targets for thieves, 164
truck weights: measurement of, at APM Terminal, 175; Virginia limits on, 40–41
Truman, Pres. Harry, 31
Trump, Pres. Donald D., 212–13
tugboats, 5, 115, 135, 139–50; articulated tug-barge combinations (ATBs), 88, 89; captains, 137, 145; Coast Guard regulations, 143; crew roster, 144–45; deckhouse colors, 240n62; integrated tug-barges (ITBs), 89; mates, 145; number used in ship docking, 126, 131–33; onboard chain of command, 146; and pollution, 160; tractor, 88; twin screw, 88
TWIC (Transport Workers Identification Credential), xiv, 198, 201
twin screw tugboats, 88. *See also* Z-drive
Twin Towers (Gillespie), x

Ukrainian government, cyberattack on, 213
underreporting of Port Newark theft, 163–64
unemployment, Great Depression, federal mitigation of, 20
unionization, decline in, 215
United States Coast Guard Forum, 93
USCGC *Elm,* 99
USCGC *Juniper,* 97
USCGC *Katherine Walker* (WLM-552), 93–103, *94, 96, 100,* 100–101, *102*
USCGC *Resolute,* 97
USCGC *Sorrel,* 105–6U.S. Lighthouse Service General Depot, 78
U.S. Coast Guard. *See* Coast Guard
U.S. Maritime Commission, 31, 43, 50, 62, 80–81
U.S. Merchant Marine Academy, 51
U.S. Navy, leases land and warehouses to Port Newark, 29
USNS *Comfort,* in coronavirus response efforts, 122
USS *Bronstein,* 23
USS *Eldridge,* 23
USS *Kearsarge,* 47–48
USS *New York* (LPD-21), 147
U.S. Steel Corporation, 22
U.S. Supreme Court, establishes NY/NJ port border, ix
U.S. Transportation Security Administration (TSA), xiv

VACIS (Vehicle and Cargo Inspection System), 185–86
Van Albert, Ivan, 242n6
Vancouver, 203
Vehicle and Cargo Inspection System (VACIS), 185–86
vermiculite, 30
Verrazzano, Giovanni da, 70
Verrazzano Narrows Bridge, 81–82, *82,* 143–44, 239–40n42

Vessel Traffic Service (VTS), 73, 95, 110–13; Sector New York, 85
vestry, in Episcopal parish finances, 206
Victory, John F., 21
Victory ships, 31, 32
Vietnam war: and containerized supply shipping, 4; and expansion of Cam Ranh Bay harbor, 64–68
Virginia, truck weight limits within, 40–41
VLCS (very large container ship), 127
Vries, Rene de, 226

Wadsworth, Gen. James S., 83
Walker, John, 95
Walker, Katherine, 94
Wanderer (launch), 116
Wan Hai 501, container ship, 200
War Assets Administration, 29
Ward, Mary Eurana, 14
warehouse stops, and theft, 164
watch standers, 113
Waterfront Commission of New York Harbor, 36
Waterman Steamship Corporation, 34–35
water side, 173
Weeden, Kim, 136
weight: accuracy legislation, 183–84; of container, 182–83; of imported containers, 183; limits for cranes, 182; truck, 40–41, 175
West Bank Lighthouse, 77, *77*, 78
wetlands, converted to Newark port zone, 10
Weyerhaeuser Timber Company, 18–19

wharfage, 62
wharf cranes, 177
Whittier, Alaska, port of, 63
Wi-Fi, free, at International Seafarers' Center, 197
Willey, Master Chief Petty Officer Jason, 104–6, 107, 109
William J. Romer, pilot boat, 76
Wilson, Edith (Mrs. Woodrow), 14
Windley, Carlye, 194
wind speed, limiting yard crane use, 179
World Trade Week, 1950, 30
World Wars, Port Newark's role during, 11–14, 22–24
Wowkanech, Charles, 121
Wriston, Walter, 3, 45–46

Xiamen Ocean Gate Terminal, China, 216

Yankee (launch), 116
Yantian, China, 203
Yarborough, Jim, 163
yard, container, 173, 178–79
Yokohama, 203

Z-drive, 5, 141, 144; on buoy tenders, 94
Zorovich, Capt. Simon, 122–30, *123*, 131, *131*, 135, 136, 138
Zorovich, Daniel, 135, 137–43, *142*; on greenhorns, 146; on tugboat job satisfaction, 147–48
Zorovich, Samuel, 137
Zorovich, Yakov "Jack," 136, 137, 138

About the Author

ANGUS KRESS GILLESPIE teaches American studies at Rutgers University in New Brunswick, New Jersey. A Fulbright professor and a *New York Times* best-selling author, he has written on subjects ranging from skyscrapers to superhighways. He is the author of *Twin Towers: The Life of New York City's World Trade Center* and the coauthor of *Looking for America on the New Jersey Turnpike*, both published by Rutgers University Press.